Fermented Dairy of Central Asia

Fermented Dairy

— OF —

Central Asia

Simi Rezai-Ghassemi

SHEFFIELD UK   BRISTOL CT

Published by Equinox Publishing Ltd.

UK: Office 415, The Workstation, 15 Paternoster Row, Sheffield, South Yorkshire S1 2BX

USA: ISD, 70 Enterprise Drive, Bristol, CT 06010

www.equinoxpub.com

First published 2026

© Simi Rezai-Ghassemi 2026

All rights reserved. No part of this publication may be reproduced or transmitted in any form or by any means, electronic or mechanical, including photocopying, recording or any information storage or retrieval system, without prior permission in writing from the publishers.

> *Because fermentation carries inherent risks, the reader assumes full responsibility for their health and safety in choosing to make and consume fermented foods.*

British Library Cataloguing-in-Publication Data

A catalogue record for this book is available from the British Library.

ISBN-13 978 1 80050 701 2 (hardback)
978 1 80050 702 9 (paperback)
978 1 80050 703 6 (ePDF)
978 1 80050 719 7 (ePub)

Library of Congress Cataloging-in-Publication Data

Names: Rezai-Ghassemi, Simi author
Title: Fermented dairy of Central Asia / Simi Rezai-Ghassemi.
Description: Bristol, CT : Equinox Publishing Ltd, 2026. | Includes bibliographical references and index. | Summary: "The book covers the origins of well-known dairy ferments such as yoghurt, as well as lesser-known ones like qooroot. There are easy-to-follow recipes for making these ferments, suggestions for how to enjoy them, and explanations of how to use them as ingredients in food, drinks, and desserts. Alternatives for vegans or those with allergies are also provided. This book is an essential resource for anyone interested in adding simple, nutritious, gut-healthy foods to their culinary repertoire while also learning about the history and provenance of food"-- Provided by publisher.
Identifiers: LCCN 2025020295 (print) | LCCN 2025020296 (ebook) | ISBN 9781800507012 hardback | ISBN 9781800507029 paperback | ISBN 9781800507036 pdf | ISBN 9781800507197 epub
Subjects: LCSH: Fermented foods--Asia, Central | Fermented milk--Asia, Central
Classification: LCC TP371.44 .R493 2025 (print) | LCC TP371.44 (ebook) | DDC 641.4/63095--dc23/eng/20250716
LC record available at https://lccn.loc.gov/2025020295
LC ebook record available at https://lccn.loc.gov/2025020296

Typeset by Scribe Inc.

*For all who provided milk, cared, created,
and shared their knowledge and ferments*

Contents

A Note on Transliterations and Romanization ix

Introduction 1
 1. Origins 11
 2. Udder Beginnings 23
 3. Fermentation, a Microbial Marvel 42
 4. Milk's Magical Metamorphosis 62
 5. Which Came First, the Yogurt or the Pot? 65
 6. *Gooroot*: Dried Dairy Balls 77
 7. A Sour Sidekick 95
 8. Spinning, Laughing, Dancing to Her Favorite Song 98
 9. A Separation: *Su & Dan* 113
 10. Whey Ahead of You 124

Acknowledgments 147
Notes 149
Bibliography 153
Index 167

A Note on Transliterations and Romanization

Transliterating and romanizing words from an oral language presents significant challenges, particularly in the absence of an established alphabet or written texts for reference. Moreover, the vast geographical regions discussed in this book encompass numerous languages and dialects, each with its own variations in pronunciation for identical or closely related terms. I have endeavored to maintain consistency to give readers an approximate sense of how these words sound in my mother tongue, Āzari.

Introduction

It was *gismat* that I should write a book on the dairy of my heritage. Back in 2009, when I first started Simi's Kitchen, my cookery school, the very first dish I taught was *kashke bādemjān*, a classic Iranian dish made with aubergines and a fermented dairy called *kashk* in Fārsi and *gooroot* in Āzarbāijāni (from now on, Āzari). Over a decade later, in 2022, I had the chance to share this dish as an hors d'oeuvre at the Oxford Food Symposium. The following year, I wrote a paper on *gooroot* for the symposium, an endeavor that ultimately evolved into this book.

I will start by setting the scene, and in order to do so, I need to highlight that when exploring the history of food or language, it is important to remember that the concept of the nation-state is a modern one. Historically, the region this book is set in was a land with shifting boundaries, migrating populations, and transient kingdoms and cities. It is convenient to use geographical names, but we shouldn't project contemporary ideas like nation-states onto the past.

Viewing Eurasia as a unified landmass rather than dividing it into "Europe" and "Asia" also highlights the complex historical and cultural interactions that have significantly influenced global food and culture. Historian Michael Bonner states,

> Events at one end of the globe can have consequences in very distant parts. Anyone with any historical imagination may have thought of that general principle. But without an emphasis on the unity of Eurasia, the history of the world will appear to be random and perhaps even unintelligible.

In the past, Eurasia was scarcely acknowledged, but today it occupies a central place in global discussions. Even the term *Eurasia* itself emerged only in the late twentieth century. This narrative has been disrupted by a broader perspective that highlights the crucial role of the area in shaping Europe's history. You may also notice I do not mention the evocative but arguably useful shorthand *Silk Road*, which was coined in the nineteenth century by the German geographer Ferdinand von Richthofen, reflecting a very Western-centric view of the history of the area. It generally implies a linear movement of goods, particularly silk, across land in Eurasia, primarily driven by European interests.

Strictly speaking, the Caucasus (where I'm from) does not fall within the geographical concept of "Central Asia." However, the many designations applied to these lands are fluid and contested. Then there are terms like "Middle East" or "Far East," derived from their proximity to past colonial European Empires. In this book, "Central Asia" is employed more broadly to denote the expanse stretching from the western frontiers of China to the Caucasus. Likewise, it is used to refer to the interconnected linguistic and cultural spheres of Turkic, Mongolian, and Iranian peoples of Eurasia.

Such labels as Iranian, Turkic, and Mongolia are used to describe the shared linguistic, cultural, and culinary practices that have been influenced by these groups, many of which persist today. Similarly, I use modern country names like Georgia purely as geographical markers, while recognizing these lands were previously inhabited by different peoples under different names. This is partly why I do not discuss the fermented dairy traditions of the Slavic or Anatolian peoples, as they fall outside the scope of this book.

Āzerbaijān is a good example of shifting borders. The Republic of Āzerbaijān was part of the Russian Empire and later the Soviet Union until the 1990s. During the Russo-Persian Wars of the nineteenth century, the Qajar dynasty of Iran was forced to cede parts of its northwestern territories (Āzarbāijān) to Russia under the Treaties of Gulistan (1813) and Torkamanchāi (1828). From that point onward, that part of the region remained under Russian control. Following the period of glasnost and the eventual dissolution of the Soviet Union, Āzerbaijān became an independent country.

This often leads to confusion about my own origins. I am from Āzerbaijān in Iran, not the Republic. It is important to note that Āzaris are not a distinct race but an ethnic group shaped by a shared Turkic language and

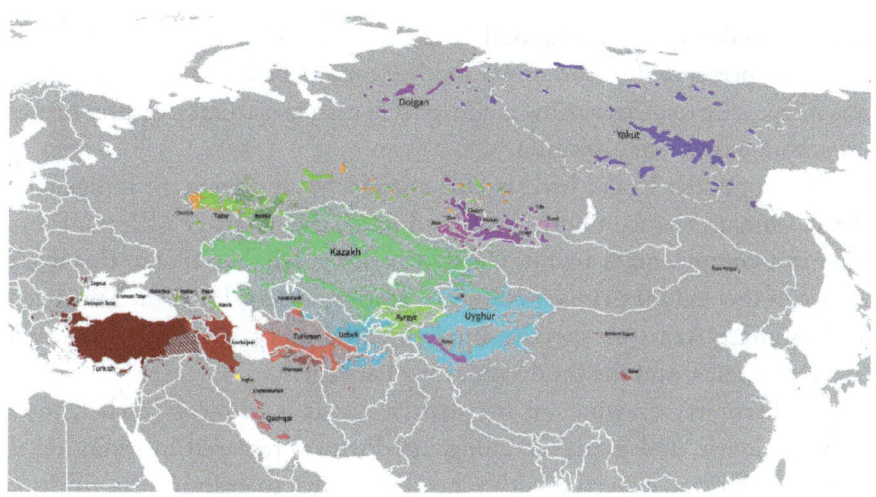

Distribution of Turkic languages. *Source:* Wikimedia.

culture. Today, there are almost double the number of Āzaris in Iran than there are in the Republic of Āzerbaijān.

Āzari is one of the many Turkic languages spoken throughout the region. Because I am from Iran, the version of *turchi* (Turkish) I speak includes many Iranian and Arabic words and pronunciations. Another important factor is that in Iran, we are not taught to read or write in Āzari; it remains solely an oral language. This is because, in the early twentieth century, there was libricide (destruction of books) along with wars and government decrees banning the use of the Āzari language. Since then, almost all the published material in Iran has been in Fārsi. I do not have access to any written material or documents; instead, I have to rely on my own exploration.

We Āzaris in Iran are bilingual, like the Welsh in the United Kingdom (I reference Wales, as I spent part of my childhood in Anglesey). While Fārsi is the official language and lingua franca of Iran, at home and in the bustling bazaars throughout Iran, we speak our mother tongue, Āzari, which is a Turkic language.

Additionally, in recent times, the influence of social media and satellite TV, particularly from Turkey (where they speak a very similar language that is easily understood by Āzaris), has led to rapid changes in our Āzari language, pronunciation, fashion, and food. This accelerated shift makes it even harder for me to find "old" Āzari terms, phrases, or sayings, particularly those related to cooking and food. This makes my work both essential and urgent to preserve the food of my mother tongue before it is lost. Even so, it is interesting to observe how language still manages to survive after brutal laws, conquest, or colonization.

The approach I take here is also shaped by my being an indigenous Āzari woman, fluent in both Āzari and Fārsi, the language of Iran. These linguistic competencies are not merely instrumental; they anchor this work in two of the region's primary cultural traditions. Here, I present a nuanced and intimate Āzari perspective on fermented dairy, shaped by my own lived experience and the foodways of my heritage.[1]

Furthermore, drawing on my experience as a cookery teacher, food anthropologist, and gardener, alongside doctoral research in evolution, I combine practical expertise with scholarly inquiry to examine the cultural, historical, and scientific dimensions of Central Asian fermented dairy traditions. Through this interdisciplinary lens, I undertake narrative-driven investigations that trace the transformation of milk from its role in ancient human practices to contemporary kitchens.

The findings in this book are based on a review of literature as well as qualitative research methods. I have incorporated insights from cultural and food studies, recipe books, travelogues, artifacts, linguistics, history, geography, ethnography, botany, and scientific and medical research. I also visit the area regularly and rely on firsthand observation and interviews. At times, I have to rely on my intuition, which feels especially relevant for a topic like this.

There were a handful of texts relating to the history of fermented dairy of the region in English. Overall, while well intentioned, some of what I read and encountered was often shaped by nationalistic perspectives, hearsay, misunderstandings, and inconsistencies.

I also connected with people from various communities—Āzaris, Kurds, Iranians, Shāhsovan, Gashgāi, Afghans, Arabs, Uzbeks, Kazakhs, Uyghurs, and others—to gather their insights and perspectives. My interpretation of their accounts is naturally shaped by my understanding and personal viewpoint. I have aimed to present findings that are well supported while recognizing that there is still so much more to uncover.

I have enjoyed looking at the etymology and names of fermented dairy foods in different languages of Central Asia. They are often descriptive, like *gooroot*, which means "dry it" in Turkic languages, like "sour dough" in

English. These words are like fossils, giving us clues about the societies they come from; it is almost like the food is speaking to us. Generally, the names for dairy products are similar and reflect the dominant culture in the area, be it Iranian, Turkic, or Mongol. Though it is the same word, each area pronounces it differently.

To illustrate this, I must admit I was quite befuddled when I first researched *gooroot*; when I was talking to people from different regions or reading about it in English, there were so many pronunciations and spellings in English texts. My Kazakh friends referred to it as *qurt*. It appears in English as *qooroot*, *kooroot*, *kurut*, *qurut*, and *kurt*, as well as many other variants. Each of these words was romanized from a local dialect or accent, either by locals writing in English or by someone interviewing a local and writing out what they heard. I, too, have romanized it the way we say it in Āzarbāijān in Iran—that is, *gooroot*—and added yet another spelling in English. This is not meant to add to the confusion, but to honor the way we say it in Āzari. None of these spellings are inherently right or wrong, but having a consistent English spelling helps make it immediately clear what is being discussed. More widely known products in the West, like kefir, though still pronounced in many ways, tend to converge on a single spelling.

Whenever I meet new people, if it is a multicultural crowd, we look for shared words for plants, ingredients, food, or experiences. Yes, there is a theme: My conversations usually start with food. In this respect, I agree with anthropologist Leo Coleman: "Nobody, not even anthropologists or other researchers, can go without food, and it should be obvious that food is somehow part of any attempt to understand social life and relationships."

The names, pronunciations, and ways the fermented dairy products are eaten in this book reflect my personal experiences, including yogurt I have tasted or witnessed being made. Each geographic region seems to have its own unique terminology, pronunciation, and of course, bacterial communities associated with its products. The fermented dairy and names discussed here are neither exhaustive nor definitive. I respect and acknowledge this diversity.

In the chapters that follow, we will be tracing the history of dairy products, some of which are the unsung heroines (yes, I deem them female) of our kitchens: yogurt, butter, and cheese, which are well known and widely enjoyed around the world. Then there are the lesser-known varieties, such as *gooroot*, *kumis*, and *ayrān*. Gooroot, which I mentioned at the beginning of the chapter, is a preserved dairy product that, in its traditional form, is dried, lightweight, and portable. She is quite a gal, being credited with fueling the Mongol conquests and sustaining cosmonauts in space, and she may even have been used as grave goods. Yet despite her fascinating history, she remains relatively unknown outside her region of origin.

Historically, the making of fermented dairy has been women's work. Women milked animals, tended fires, churned butter, and coaxed microbes into transforming milk into more stable, nourishing food. By highlighting this here, my book seeks to honor all females' contributions. Similarly, I use the correct names for animals that produce milk; too often, it is said sheep, goat, or yak's milk. In fact, milk is produced by female animals: ewes, nannies, and dri.

Even now, in traditional societies, gender roles are quite distinct but often characterized by relative equality compared to many modern societies. Generally, women oversee dairy processing, while men are engaged in herding. This division of labor highlights the collaborative nature of survival strategies, particularly in nomadic and pastoral groups. Foregrounding females, both animal and human, underscores their role, often excluded when narrated through a male-centered lens of conquest, trade, travel, or politics in this area.

Upon each pot of yogurt, each skin bag of fermented mare's milk, there are women's hands. This work has been shaped by the rhythms of milking and the seasonal cycles of yurt management. These women developed skills to sense when milk would set and when butter was ready. They adjusted to seasons, animal health, and microbial mysteries. Such knowledge was transmitted through practice rather than formal instruction. Young girls learned by assisting their mothers, absorbing embodied techniques. Dairy fermentation was women's knowledge, intuitive, intimate, domestic, and passed down matrilineally, and it was never written. In my own research and visits to villages, when I asked, "How will you know when it is ready?" or "How much of an ingredient?" more often than not, they would say, "As much as it needs" or "You will feel it."

In this book, there are a few photographs of people, and they are of the men. That is because they gave me permission to take and use their photos, but the women declined.

In most parts of Central Asia, dairy products are known as *āghārty* (white foods), which continue to be important foods, particularly during the summer months. Interestingly, I found the medieval Irish word *bánbhia*, meaning "white food," which also refers to dairy. In addition to their nutritional value, *āghārty* hold symbolic meaning and are greatly esteemed. They are believed to possess powers that could bring health, wealth, and longevity. In most of these cultures, fresh, unprocessed milk plays a role in sacred or ritualistic ceremonies; it was sprinkled at weddings, behind a traveler, or on the head of a winning horse. I witness this respect for milk with my Panjabi friends, who are uncomfortable if they have to pour spoiled milk down the drain, much like us Iranians, who will not throw bread in the bin.

There was and still is prestige attached to cultures skilled in processing dairy. *Āghārty* are still offered as a welcome food to honor guests, and as part of dowries and ornaments, for example, small pieces of dried dairy are used as beads made into necklaces, more precious and necessary than their modern equivalent of diamonds. All these practices indicate that from our earliest times, nutritious, preservable, and portable food has been of immense value.

I consider myself fortunate to have been born into a culture that continues to hold dairy foods in high regard. In my family, growing, making, and buying only the finest foods are second nature, and it is an Āzari trait. To illustrate the close connection we maintain with those who produce our food, I will share a memory from a visit to Iran in 2025. I went to pick up some milk and yogurt from my mother's local dairy shop. There, I had a nice chat and catch-up with Mr. Abdi, the shopkeeper, who asked about this book and shared stories about his family, generations of dairy farmers and artisans. For us, dairy is never "just" yogurt or cheese, nor is it anonymous. Over time, we come to know the shepherds, the producers, and the places where our food comes from. These relationships are personal, and we often recommend our trusted sources to friends and family.

To tell the story of fermented dairy in Central Asia, I begin by tracing key evolutionary moments linked to fermentation, then examine milk's properties and its natural tendency to ferment. We will see how the peoples living in vast territories of Central Asia adapted to environmental changes, be it by migration or climate change. The region has a variety of climatic zones, including shrublands, deserts, forests, mountains, and valleys. Some are arid, and others are semiarid or temperate with seasonal weather patterns. Large parts of the area became more arid after the last Ice Age. Based on fossil evidence, our ancestors lived in these areas for millennia, gradually transitioning to pastoralism, with some communities settling while others remain nomadic to this day. The domestication of horses in this area further accelerated mobility, allowing for a more rapid exchange of populations, languages, ideas, and food.

Humans may have first started using the milk of other animals out of necessity. Milk would be plentiful in the spring or summer, when the animals were lactating after giving birth. The variety of terroir described above imparts unique organoleptic qualities, encompassing taste, scent, and texture, to milk and its derivatives, imbuing each with distinctive flavors. Such variation parallels the nuanced distinctions found in winemaking, highlighting the cultural and environmental specificity of each dairy product.

Milk has a propensity to transform. The naturally occurring bacteria in milk cause it to ferment, making it a great substrate for fermentation. The seemingly mysterious transformation of unstable raw milk into durable fermented products relies on the synergistic actions of bacteria, yeast, and fungi. Not only does fermentation change the flavor and texture of milk, but it also makes the milk easier to digest for lactose intolerant adults. By happenstance, we were consuming a food that we now know is great for our bodies, brains, and emotions. Some even argue it was fermentation, not fire, that led to brain growth, ultimately shaping our dominance as a species.

There are a host of proven benefits to consuming fermented foods in general and particularly fermented dairy, such as probiotics, and many other claims are yet to be proven. After we started deliberately fermenting dairy, our biotechnologist ancestors learned to manipulate various bacteria and yeasts for many applications of fermentation beyond food.

Fermented dairy is mild when fresh, with a flavor that develops over time, bringing a robust and, in some cases, "animally funk" to the taste buds. Travelers in this area who have partaken in some traditional nomadic dairy ferments have described them as the essence of the animal, or as if you were eating or drinking the barnyard! Besides personal preference or genetics, sometimes our reticence or desire for some foods is culturally informed. A food that a whole culture withdraws from might well have been deemed unsafe by their forebears.

In each chapter, I also examine the health benefits and risks of dairy products. In the past, people used their senses of smell, taste, and intuition to decide whether to eat or discard food; now we rely on use-by and sell-by dates, which outsource this decision. Some labels on packaging warn consumers about the levels of sugar or fat in food. Increasingly, we are advised to read the ingredient lists and to put the product down and walk away if we see ingredients we would not find in a domestic kitchen.[2] However, in the UK, people are cooking less than ever before; it may be that those who do not cook may not know what to look for.

Modern mass-produced "foods," which are often described as tempting, are addictive, synthetic, and convenient. They are produced with cheap materials and are typically in disposable packaging. These have contributed to an increasing detachment from whole foods and are polluting our bodies and the environment. Similarly, the animals that give us milk and the milk itself, or should I say "mass-produced white stuff," the base of all dairy products, have changed, and as a result, the products we consume today are not the same as those our ancestors once had.

I have always been fascinated by science and technology, and I am in awe of what we have accomplished. However, I believe this knowledge and technology should be used for the betterment of our planet and all life on it. It puzzles me that we have been led to believe that natural, raw, healthy milk from well-kept animals in clean environments is more dangerous than sterile, synthetic, genetically modified "milk" produced in laboratories by people in white coats. With growing interest in novel ingredients, fermented foods, longevity, and the gut-brain connection, I aim to introduce you to, and encourage you to seek out, high-quality milk and fermented dairy products, which are nutritious, ancient foods that humans may have coevolved with.

I find it encouraging to see many Central Asian dairy products listed as part of the Slow Food Foundation's Ark of Taste initiative. Of the 6,439 entries, 549 are dairy products from Central Asia.[3] This highlights the diversity of these similar yet distinct dairy products. There is also UNESCO's list of Intangible Cultural Heritage, which focuses on nonphysical cultural elements, like knowledge, skills, and rituals that define and shape cultural identities. The heritage list includes a wide range of food-related traditions, examples of which include baguettes from France, *yarpākh dolması* (stuffed vine leaves) from Āzarbāijān, and *airag*, a fermented dairy drink from Mongolia. This recognition demonstrates the significant pride that communities take in their local variations of foods, especially the dairy products of the region and their importance to our human food history.

What follows is a basic overview of the story of Central Asia's most popular fermented dairy products—a taster, if you will. It is a record of humans, animals, milk, and transformation, a considered and careful account of the knowledge, practices, languages, and cultures embedded in these foods. Above all, this work represents an urgent and vital effort to preserve and document the histories and foodways of this region, often considered the cradle of food, where time-honored traditions have nourished communities for millennia and laid the foundations of our shared global culinary heritage.

1
Origins

The study of fermented dairy offers a fascinating lens through which to view the history of our evolution. While precise knowledge about human evolution remains limited, advancements in science and technology and the application of artificial intelligence have allowed us to infer patterns and trends from existing findings. While in this book I focus on large parts of Eurasia, that does not mean comparable progress did not occur elsewhere; each location has its own unique regional variations, and new finds and excavations are giving us more details to help piece together our human history. This chapter will briefly focus on the key factors that shaped our journey toward becoming both consumers and producers of fermented dairy. This exploration will travel through time to emphasize essential elements, though not necessarily following a linear progression. I have also taken the liberty of making up short, speculative scenarios that may or may not align with future findings about human development leading to the fermenting of dairy.

We believe the earliest hominins were predominantly vegetarian, coming down from the trees about four million years ago to forage on the forest floor and adapting to bipedalism in the process. This gradual transition over millennia brought about significant morphological changes, including adaptations to the rib cage, breathing mechanisms, and coordinated limb movements, which also freed the hands for diverse uses. The physical adjustments, alongside changes to brain size, were responses to the challenges and opportunities of a new environment. Over time, we adapted not only to varied ecosystems and seasonal shifts but also to abrupt or gradual environmental changes, whether geological or climate related.

In our eastern expansion out of Africa, our ancestors went across to the Caucasus and beyond. The Dmanisi archaeological site in Georgia is where fossil skeletons of *Homo erectus georgicus* have been found, the oldest hominid found outside Africa, dating from almost two million years ago. Roughly a thousand kilometers to the south are the limestone caves and valleys of the Zagros Mountains, a vast range stretching from Iran into northern Iraq and southeastern Turkey. Archaeological finds here indicate these caves provided shelter to later travelers moving through the region, though the mountains themselves may have acted as natural barriers to human migration. The region's varied landscape, including high mountains and fertile grasslands, also supported abundant flora and fauna, offering a wealth of resources. Early humans in this area likely had access to water and a wide range of foods, such as insects, fish, birds, animals, nuts, seeds, berries, fruit, tubers, vegetables, and herbs. They would have initially focused on the easiest food sources to gather, hunt, or scavenge. Trial and error, often fatal, shaped their dietary practices, a theme that recurs throughout our evolutionary history.

Our evolving olfactory system played a significant role in helping us adapt to dietary shifts. Humans possess nearly four hundred olfactory receptor genes, one of the largest gene families in our genome, and our most abundant taste receptors detect bitterness. This sensitive system warned hominins of toxins and helped fight bacteria, delivering a dual function of protection and immunity. Over time, the evolution of taste and smell receptors not only safeguarded them but also influenced their dietary preferences, shaping their evolving relationship with food. In modern life, though we rely on these receptors less, this sensitivity has persisted. The olfactory system has a direct pathway to where memories reside in our brain. We have all experienced how a scent can transport us in a nostalgic way; the nose and brain are finely calibrated to each other.

As previously mentioned, we were roaming over a range of terrains and living in larger social groups, both of which required a larger brain necessary for processing more complex social skills, spatial memory, and navigation. Natural selection favored those with receptors attuned to identifying plant toxins in new environments and remembering where seasonal food could be found. Those with the most effective olfactory receptors survived, reproduced, and passed on their advantageous genes.

Hominins were omnivores, likely consuming a wide variety of foods, including overripe, fermented, or moldy fruit, arguably the first forms of "processed food" in human diets. Fossil endocasts from this period reflect changes in our neuroplastic brain structures aligning with such dietary shifts. Bryant, Hansen, and Hecht, in their paper "Fermentation Technology as a Driver of Human Brain Expansion," published in 2023, theorize that the inclusion of fermented foods spurred significant physical and consequent cognitive development. They emphasize that the consumption of externally fermented foods rather than cooked food may have been the main reason for our evolutionary success. They reason that it may have been at the same time that early humans were consuming fermented foods that the colon, once the main organ for internal fermentation, reduced in size. This alteration allowed more calories to be reallocated to brain development, resulting in a growth in brain size and a reduction in gut size over time. Our brain is energy hungry and uses ten times more energy by weight than any other part of the body. This growing organ needed all the food it could get.

Now modern scientific research is revealing the health benefits of fermented foods, including enhanced digestion, improved gut biome health, better nutrient absorption, and increased immunity, to name a few. Fermented foods spontaneously occur in nature, and they provide an easier-to-digest food, which reduces the energy required for digestion. These characteristics of fermented foods likely made early humans stronger and more resilient. According to the theory discussed above, consuming fermented foods, together with changes in our behavior and social structures, led to the adaptations to our brain that are argued to be the reason for the ascent of man.[1]

Increases in brain size may have led to us using tools; stone tools dating back 3.5 million years have been found. Maybe some of them were used to access a wider variety of foods. Small rocks might have been employed to crack open nuts or shellfish, sticks to dig out tubers, and larger stones to mash or grind tougher foods. Initially, our ancestors might have used such tools opportunistically, much like some modern primates and other animals such as otters. Speaking of which, it was thought that otters used a rock once and then found a new one as they needed them, but in recent times, scientists have discovered some keep a "favorite rock." This and food are stored in the loose skin "pockets" under their arms.

From as far back as three million years ago, there is evidence that we started to retain tools, signaling a shift in behavior. Over time, we modified them to be sharper or pointier, eventually leading to the development of specialized tools for different tasks.

By 1.7 million years ago, humans were intentionally flaking rocks to create sharp edges, enabling them to cut, process skins, and hunt, which significantly improved their ability to adapt to the diverse climates and environments they found themselves in. These more advanced tools designed for specific tasks indicate purposeful behavior and the ability to anticipate needs. Intriguingly, most discovered tools appear to have been fashioned for right-hand use.

It is also likely that we created tools from plants or feathers, though these are rarely preserved due to decomposition. As a result, our understanding of early tool use is incomplete and skewed toward durable materials like stone. This limitation suggests that tool use may have begun much earlier than the current archaeological records indicate. For instance, evidence from 1.5 million years ago in Spain reveals that someone used a wooden implement to remove food stuck between their teeth. This is a very rare occurrence, and it is only because we now have the scientific means to analyze such finds that we know this. It is intriguing to think that humans may have used toothpicks for practical purposes or even aesthetic reasons. Such discoveries offer a glimpse into the behavior, resourcefulness, and I would argue, similarities between us and our early ancestors.

The evolution of surviving hominins significantly influenced many aspects of our behavior, particularly the development of communication skills, which could have led to teaching and knowledge sharing. Communications likely began with vocalizations, body language, and gestures, eventually culminating in the development of a spoken language. Approximately two million years ago, during a period of increased brain size, early humans may have communicated using physical gestures, bodily adornments, and sounds, potentially imitating birds or other animals. They may have even used natural materials like plants, feathers, or pigments for symbolic communication. These early forms of expression likely enabled them to convey threats, signal sexual availability, articulate needs, and share knowledge with others.

On a more instinctive level, we humans have always been attuned to the energy around us, whether from trees or other people. This intuitive sensitivity may have facilitated shared experiences during moments of threat or joy, fostering the development of consistent gestures, vocal sounds, or words to express those emotions. Subfields of paleontology, such as paleobiology, paleoneurobiology, and paleocognition, are seeking to uncover the origins of language and other cognitive behaviors by studying fossils, which could offer valuable insights into the evolution of human communication.

Communication was vital when seeking food, whether it was plants or animals, especially large prey. It turns out contrary to the hunter-gatherer archetype, early humans were likely scavengers before evolving into proficient hunters. Butchery marks on animal bones dating back 2.4 million years suggest that they scavenged meat from kills made by larger predators. We observe modern parallels in the Maasai of Kenya, who boldly approach lions after a kill to claim parts of the carcass.[2] By using basic tools, group coordination, and communication skills, the Maasai demonstrate behaviors that may have been similar to those of our ancestors.

> **Scenario 1**
> Hominins, like other animals, probably drank water directly from streams and rivers using their mouths. At some point, an individual may have used cupped hands to drink, and this behavior was observed and imitated by others. Over time, the concept of containing water to drink became ingrained in their behavior. This act likely evolved further, with someone innovating and using hollowed-out materials such as wood, plants, or shells to scoop and hold things. Eventually, when humans began to use the blood or milk of other animals, they discovered that these, too, could be held and stored in containers. These types of inventions and innovations added to our collective knowledge; as illustrated above, one such transformative innovation was container technology.

I would like us to stop here and just think for a moment about what a difference that would have made for our ancestors. Just thinking about comestibles, here was the ability to contain, store, or transport food or drink rather than eat or drink in the moment where it was found. To have only known a world where you ate or drank things where you found them and returned to the same place to get more and then moving to using containers to take what you could with you to eat later by yourself or share is quite remarkable to me.

Having transitioned to bipedalism, early humans had lost some of the natural bodily folds that they may have used to transport items. Early containers, likely made from delicate organic materials now lost to time, may have included hollowed-out plants, plant fibers, and animal skins. These vessels enabled the transport of babies, water, food, tools (like a preferred rock for mashing or cracking), fire-making kits, and (maybe without realizing it) fermentation starters. These types of behaviors indicate foresight and planning. Evidence for this includes shells discovered far inland from coastal areas, suggesting deliberate transport. Shells, while often associated with adornment, were also used for practical purposes such as scraping and cutting.

Container technology was a major leap forward for early humans. The ability to carry water away from its source, collect food for later consumption, and share resources with others was transformative. This may have also enabled the storage and transport of food that could be preserved for longer periods. Containers also introduced the broader concept of storage, since containers could be put in a cave or tree hollow. Since fermentation occurs naturally, with minimal equipment and specialized knowledge, this may further support the theory that fermentation played a central role in human advancement.

Eventually, if these vessels were used for containing milk, then it would have fermented. Regardless of the animal from which milk is sourced, fermentation begins soon after the milk leaves the animal's body. Depending on the microorganisms present in these environments, stored foods could become either poisonous or more palatable. These mysterious outcomes likely led to the development of superstitions or belief in magic, possibly related to what the milk was stored in and where. These notions may have been passed down through generations as part of cultural memory. However, this is an area of cutting-edge research, and further evidence is needed to fully understand these practices.

Even today, natural vessels are used all over the world. For example, gourds are used as containers, and in some cultures the type or design of gourd indicates the tribe, age, or gender of the owner. They are still used for containing valuable items, including food, drink, and personal possessions. In some cases, smaller gourds are used for storing sensitive items, such as parts of the body after certain rituals or medical practices!

In our evolutionary story, slowly our skills and capabilities became more sophisticated. Two million years ago, hominids started building and using shelters, and there is evidence of their encounters with fire. Despite having samples of charcoal, baked sediments, and burnt bone from the time, we are still not sure if this was a purposeful act or an accident, so we cannot say yet whether we cooked with or used fire deliberately.

The use of fire requires a diverse set of skills executed in a coordinated and sequential manner. Naturally occurring fires, sparked by events such as lightning strikes, wildfires, or the ignition of hydrocarbons from underground, would have been familiar to our ancestors living in this region with large amounts of underground gas and oil. They would have also had a sense of fire's dangers, its distinct smell, and how the environment was affected by it. It is likely they would have scavenged in the aftermath of these natural fires, consuming food found at the edges of the burned areas, such as cooked eggs, plants, or animals.

Studies on chimpanzees in Senegal reveal that they can predict the behavior of wildfires, determining when and how to move to avoid danger. After the fires subside, they forage in the burned areas. However, unlike humans, they do not utilize fire itself. Pyrotechnology, the ability to harness and manipulate fire, is a uniquely human skill, setting us apart from even our closest evolutionary relatives. Ash and charcoal remains from eight hundred thousand years ago provide better evidence to suggest that humans developed the ability to harness fire, marking a pivotal moment in our evolutionary history. Fire became an essential tool; maybe to begin with it was used for protection, warmth, and light and later for cooking, toolmaking, rituals and ceremonies. As such, fires likely became central to social life, serving as gathering points for groups and fostering interaction and community.

Primatologist Richard Wrangham and many others suggest that it was cooking that led to bigger brain size in hominids. The idea is that our larger brain and smaller digestive system evolved because we started eating cooked rather than raw food. Cooking not only makes food easier to chew and digest, contributing to a more nutritious and bioavailable diet, but also eliminates pathogens. However, as previously mentioned, there is limited evidence of cooking before eight hundred thousand years ago, suggesting this theory may be more relevant from that point onward. While humans eventually tamed or harnessed fire, the precise timing and location of this breakthrough remain uncertain. There is stronger evidence of fire used deliberately by the discovery of structured hearths from four hundred thousand years ago. Their size and number are influenced by factors such as the availability of nearby fuel, the climate, and their intended use.

Moving on now to our relationship with other animals in our environment, it is believed that gray wolves and maybe foxes first came into human lives approximately fifteen thousand years ago, likely becoming the ancestors of modern dogs. Initially, these animals would have approached human settlements for food scraps, establishing a symbiotic relationship in which they offered protection in exchange for food. Over time, humans trained and bred them for purposes such as hunting and other tasks.

As humans realized that long hunting and scavenging trips were time-consuming and energy intensive, they began trapping and breeding animals that could be herded. This shift provided a steady, renewable source of food without the same level of risk or energy expenditure. Archaeological evidence from around twelve thousand years ago also indicates that sheep, goats, and other caprines adapted to semiarid steppe environments were managed and herded in this part of the world.

To begin with, these animals were likely treated as game and managed for food, being consumed after natural death or slaughter. Although there is some debate surrounding this theory proposed by archaeologist Andrew Sherratt, who suggested that early herded animals were first used as "deadstock"—meaning they were consumed for their meat and blood after death—it is a compelling theory and plausible. According to Sherratt, it was only later that humans realized these animals could be maintained as livestock, providing resources such as wool, transport, traction, and milk.

Our ancestors likely observed how herded animals fed their young with milk, just as we did, leading them to obtain some for themselves. Initially, this was probably done out of necessity during times of famine or hardship, but over time, it became a more regular practice. Some crude and likely unreliable theories suggest that human–animal interactions of a sexual nature may have played a role in the domestication process. Regardless of how it began, humans started milking animals. In order for us to milk them, the animals would have needed to be approachable, docile, and capable of producing more milk than was needed by their offspring.

Having collected the milk, our ancestors faced a problem. We can only assume, based on what we know now, that they too lacked the enzymes necessary to digest lactose. Although it may have caused discomfort, it was not fatal. We have seen that once milk is out of the animal, if not kept cold, it will naturally ferment; therefore, we can assume most likely they were drinking or eating fermented milk. Consuming fermented milk, by contrast, alleviates some of the digestive issues and offers additional benefits, making it a valuable new food source.

This raises intriguing questions: Did our ancestors realize the difference between raw and fermented milk? Why are some people today lactose intolerant while others are lactase persistent, able to digest milk without issue? Is this tied to inherited DNA from our ancestors? It is often suggested that cultures relying on fermented dairy became lactose intolerant, while those consuming fresh milk evolved lactase persistence. Could it be that drinking raw milk led to the reactivation of genes coding for lactase production, allowing populations to digest lactose through epigenetic mechanisms? Or, conversely, did populations that primarily consumed fermented dairy lose the need for lactase production, eventually becoming lactose intolerant as the gene became inactive over generations? These questions remain open for exploration.

Before I move on, we also ought to bear in mind that the discoveries in nearby Göbekli Tepe and Karahan Tepe in southern Turkey are considered the earliest known sites built specifically for human gatherings. Göbekli Tepe is thought to have been a type of temple, while Karahan Tepe, an older and larger site, served both domestic and ritualistic purposes. The region's abundance of wild cereals; grazing animals like sheep, goats, and gazelles; and seasonal migratory species likely drew humans to this area in the first place. The presence of pistachio and almond trees, evidenced by charcoal remains, provided resources for food, fire making, and other development needs.

These sites date back approximately 11,500 to 10,200 years ago, respectively, suggesting that the knowledge and skills required to construct such monumental structures likely developed much earlier. Although researchers have not yet determined how the civilizations that built these sites sustained themselves, further discoveries could push back the timeline for the first human foods and dairying beyond our current estimates.

Generally, humans would have made their dwellings near water, gathering around springs or clean sources of running water. To the best of our knowledge so far, following the end of the last Ice Age, large parts of the region became more arid and unsuitable for agriculture but suitable for pastoral life. Humans moved seasonally with their herds in search of better pasture while also relying on foraging, hunting, and scavenging.

To the south of the Caucasus all the way to Anatolia, what is now known as the "Fertile Crescent," grew warmer and wetter, supporting the growth of grasses.[3] Current knowledge indicates that around eleven thousand years ago, some of the people moved to this area and began to transition from a nomadic, hunting-and-foraging lifestyle to agropastoral settlements. These early communities utilized local resources to grow crops, construct dwellings, and create tools, significantly altering the local ecology. It is still being researched, but it is thought that relying on a few cultivated crops, rather than a diverse foraged diet, led to poorer health and, in bad years, more sickness and death.

Settled living also brought about other new challenges as well, with domesticated animals—including goats, sheep, and pigs—being kept alongside humans. Not only did these animals contribute to soil degradation, but also the increased presence of manure attracted vermin and associated pathogens. Some bones found at this time confirm disease and poorer health.

Turning our attention to evidence of dairying in the area, preserved human teeth found here were analyzed and have revealed that dairy ferments made with ewe's milk were consumed over nine thousand years ago in the Caucasus. Analysis of dairy proteins has given results consistent with previous lipid analysis, which showed that milk was processed into dairy products. In more recent times, not only can archaeologists establish what vessels were used for, but they can also identify the type of animal, plant, or food contained in the vessel. Some lipids and protein molecules, which can last for thousands of years in cold climates, are beginning to help us better understand prehistoric food and cookery practices. Proteins carry a lot of molecular information that can be analyzed to establish a great deal of detail about them. In this case, the proteins are telling us about the processes used to make dairy products. These technologies, along with emerging methods, allow us to identify what was stored or cooked in the pottery vessels and sherds that archaeologists uncover.

We get even more insights into dairy processing from other agricultural settlements like Çatalhöyük in southern Turkey, which is dated to around seven thousand years ago. Discoveries here indicate that its inhabitants were refining their agricultural techniques and advancing the domestication of animals. The addition of grain to the diet spurred the invention of new tools and the development of advanced grinding stones. Grains also became a key ingredient in various foods, including bread, a product that underscores the early role of fermentation in food preparation and preservation.

What is more, at Çatalhöyük, pottery served a broader range of purposes, extending beyond storage and cooking to include the processing of agricultural produce. Analysis of whey protein on pottery sherds found throughout the region tells us that ewe's and nanny's (female goat's) milk was being processed into curds and whey using beeswax to waterproof sieves. This indicates a method of separating curds and whey, or cheese making.

Later discoveries of dairy product residue in 4,500-year-old pottery at the Harappan site in Pakistan, along with numerous containers, strainers, and graters, suggest that large-scale dairy production was occurring at the time. These findings are significant, as the residue found was not cheese made with rennet; rather, they were curds and whey produced as a result of the natural souring or curdling of milk. DNA analysis at the Harappan site also reveals that buffalo cow's milk was used for dairy production, with archaeobotanists identifying millet, a common crop in the area, as part of the animal's diet. Additionally, sheep and goat remains show signs that the animals were also eating grass and leaves. All these pieces of the jigsaw puzzle help us get an idea of the local flora, animals, and combinations of foods consumed by our ancestors and allow us to reconstruct the food of the period.

Another, to me, enchanting find is a necklace threaded with chunks of dairy. It was discovered in the Taklamakan Desert in northwest China. The preserved remains of the "Beauty of Xiaohe" were uncovered in 2003. She was buried around 3,800 years ago, wearing a felted wool hat, string skirt, and fur-lined leather boots. Along with her attire, she had several other items, including what was described as a "cheese necklace." Researchers found that the "cheese" was made using bacteria and yeast. I reached out to them in 2023, inquiring whether it could have been a form of *chortān* or *gooroot*, which are two forms of dried yogurt and whey, or possibly even kefir, as they didn't find evidence of rennet being used. In this book I will use the word *cheese* when a curd is formed by use of rennet to distinguish it from other curd products; that is why I have put *cheese* in quotation marks. As we go to publication, I await their reply. It seems plausible that it was dried clabber, *chortān* or *gooroot*. These are all nutritious and important foods, which may have been used for sustenance during travel or migration and might have also been included in the journey into the afterlife.

There is still a custom, especially in nomadic and traditional families in Central Asia, where brides are given necklaces threaded with precious items to take to their new homes, sometimes in new lands. These items could be amulets, seeds, and even dairy starters.

In the thirteenth century, the region was dominated by the Mongols, and it is documented that they followed distinct seasonal diets: a summer diet of dairy-based *āghārty* (white food) and a winter diet of meat, known as "red food." Meat consumption in colder months persisted until refrigeration became widespread. My mother shared that in her village, up until the 1960s, butchers were closed in the summer months. The butchers would have only been called upon to slaughter or dress animals during festivals and celebrations in the summer. Meat for summer use was preserved as *govurma*, the result of a method of preserving mutton in animal fat and storing it in cool cellars. This technique is now rare, with only one of my mother's friends still making *govurmā*, a tradition that is gradually dying out.

In the upcoming chapters, I will look at the animals chosen for milk production, milk, and the most common fermented dairy products. These foods have played a significant role in shaping the history of the world and have become symbols of cultural identity in Central Asia.

Some scholars argue that civilization developed from human empathy, altruism, and the care we give to one another, an idea often associated with anthropologist Margaret Mead. Others suggest that rituals, the control of nature, the harnessing of fire, and the consumption of externally fermented foods were key drivers in our evolution and subsequent dominance. Current advancements in genetic technology are helping to reveal how these physical and behavioral adaptations to various environments have shaped us. The evolutionary theories of biology proposed by the polymath al-Jahiz and naturalists Alfred Russel Wallace and Charles Darwin, along with psychologist Susan Blackmore's meme-based accounts of cultural evolution, lend support to these ideas.

2

Udder Beginnings

As mentioned in the previous chapter, historically, humans transitioned from foraging, hunting, and scavenging to domesticating animals and cultivating plants. This was, in many ways, the beginning of our departure from nature and into a more structured way of life. Unlike a natural setting, gardens and farms represent deliberate human intervention, defined by boundaries, planning, and maintenance. This period of starting to manage nature is often heralded as the birth of civilization, though the exact definition of *civilization* remains a subject for debate. I have often wondered, if that is the case, then why do Eurocentric narratives associate civilization primarily with Greece and Rome?

In this chapter, I will explore the animals that provide milk, discussing what milk is and how it is obtained and prepared for consumption. I will also examine the cultural importance of both the animals and their milk and dairy products in the lives of the people of Central Asia.

Among the many ways we shaped nature to suit our needs was to herd sheep or goats (it is unclear which species came first), which transformed human societies. For example, those in regions with year-round pasture may have remained stationary, while those without may have had to migrate seasonally, most likely following the animals to greener pastures. Eventually, entire communities, along with their flocks, would move to and from summer or winter pastures and back again, a practice known as pastoral nomadism. Later, when humans started living in one place, some also practiced transhumance, where a few members of the tribe would travel with the herd. This lifestyle is often romanticized in art, typically depicting shepherds, usually men or boys, looking after their flocks in idyllic landscapes.

Though it might be so at times, for the most part, it is hard work and brutal in bad weather or when crossing difficult terrain.

Let us now look at milk, which starts with sex; mammals produce milk solely to feed their offspring. In the past, there would have been milk gluts coinciding with lambing, or whenever female animals were giving birth. As mentioned in the previous chapter, throughout this book, I will make a point to acknowledge the contributions of female animals to this story. While female cows receive recognition, species like camels and buffaloes, whose females are also called cows, deserve distinct credit as well. Additionally, I will highlight other female animals—sows (female pigs), ewes (female sheep), nannies (female goats), jennies (female donkeys), mares (female horses), and dri (female yaks and reindeer)—to emphasize their contributions. A small nudge, if you will, in pushing back against dominant patriarchal narratives.

After giving birth, female mammals produce the largest quantity of milk, and the amount naturally decreases after their young are weaned. It would have been at this time, when animals have a lot of milk, that some of it was collected for human use. Moreover, lactation periods vary among species: Ewes lactate for eighty to one hundred days and camels for up to a year, while modern dairy cows are bred to produce milk for ten months of the year.

I find it astonishing how plant matter like grass is transformed into milk. Of the mammals, ruminants such as goats, sheep, and cows process plant matter over a long period of time, chewing it repeatedly, which is how they get the name ruminant. First, they collect the plant matter by cutting it off with their teeth and swallowing it. Then they rest and regurgitate the cud, which is the semiprocessed food returned to the mouth to be chewed again. Oddly, cows tend to lean left while chewing their cud! This process involves the physical action of grinding with their teeth and a chemical reaction with their saliva. Afterward, the plant matter goes into another of their four-chambered stomachs, where bacteria and enzymes break it down further into carbohydrates and amino acids. Some of this is absorbed by the body to maintain the animal, and in female animals, some of it is recombined into fatty dairy, which is excreted from their udders as milk.

Milk is also associated with nursing mothers and newborns; it is our very first food. The initial milk produced by mammals is called colostrum, known as *chāllā* in Turkic languages and *āghuz* in Iranian languages.

Colostrum is nutrient dense, is rich in growth factors and antibodies, and is believed to also seed the gut microbiome necessary for building resilience in newborns. Outside the sterile environment of the womb, infants face exposure to pathogens and the challenge of gravity, making this essential nourishment when they are most vulnerable.

In many cultures, animal colostrum holds a special status and is often shared as a prized food with family and guests. In parts of Āzarbāijān, for example, during February, the yellow, thick colostrum from ewes and nannies, called *chāllā*, is slowly stirred into milk until it becomes uniform. The process of stirring and the resulting food are both known as *boolamā*. It is a seasonal treat and can be enjoyed plain or sweetened with honey or grape molasses. Many countries use colostrum in different ways; for example, *junnu* and *kharvas* are two of the many different names for an Indian sweet dish made using cow or buffalo cow colostrum. It is also made into cheese in some areas.

Another type of *"boolamā"* I came across was made in the autumn in a remote village of northern Āzarbāijān. I couldn't understand how they had colostrum at that time of year, only to later discover that it was not made with actual colostrum. It was called *boolamā* because the color and texture were very similar. In the autumn, ewes graze on dried grass and other plants, and the lack of moisture makes their milk yellow. Furthermore, it is thicker due to the lack of moisture in these plants, and that is why they call it *boolamā*.

As we have just seen, many factors influence the milk's taste, fat content, color, and organoleptic qualities. Some animals, due to their biology, can produce fattier dairy. The fat content significantly affects the dairy products—for instance, whether they yield creamy cheese or lighter yogurt. Milk has a remarkable capacity to transform, spontaneously or through deliberate processes, into an array of bioavailable and nutritionally dense foods, which I will discuss in the following chapter.

Overall, across the globe, why humans predominantly consumed milk from certain species rather than others offers insights into the coevolution of human societies and their choice of animals. The pattern of milk consumption we see today emerged through a complex interplay of biological, geographical, and cultural factors that shaped our relationships with different mammals over millennia.

Central Asia was rich in mammals, some of which are now extinct, like the woolly mammoth and auroch, or near extinction, like the saiga antelope and onager. There were many mammals: bears, marmots, wildcats, wolves, and boar, to name some. Animals such as sheep, goats, horses, camels, buffalo, and yaks were also well suited to the region's terrain and climate, thriving in the arid and semiarid conditions. These were some of the many mammals that offered potential milk sources. However, practicality dictated the animals humans chose to milk. We are unsure why certain mammals were chosen, but medium-sized, approachable herbivores with accessible udders seem to have been preferred over carnivores or omnivores.

Let us consider first why we might not have chosen to consume milk from our oldest and closest companions, wolves or dogs: Why weren't they selected for milk production? She-wolves and bitches typically produce milk only for short periods and in quantities suited specifically to their pups' needs. Additionally, their litter-based reproduction pattern means milk production is sporadic rather than continuous, making them unreliable as a consistent milk source.

Wild boars and pigs are closely related to each other and are native to Eurasia. They are called *douuz/domuz* in Turkic languages and *khook* in regions influenced by Iranian languages. These nocturnal omnivores have been domesticated independently across various parts of the world, beginning about nine thousand years ago. They produce numerous offspring per litter, and their milk-production systems have evolved to meet the needs of multiple offspring. Sows are also known for their intelligence and become particularly unapproachable and aggressive when caring for boarlets or piglets. Being omnivorous and unapproachable may have been the two main reasons we did not use their milk. Of course in the last few millennia, religious traditions in the region have prohibited the consumption of this animal and its milk. Sow's milk is high in protein and fat, making it potentially excellent for cheese making.[1] Apparently if we really wanted to milk a sow and could get to the teat, it turns out that human breast pumps are a good fit. The mind boggles at how this was found out.

Now, as to the animals we chose to use for milk—ewes, nannies, camel and buffalo cows, mares, and dri—they all share several characteristics that make them better candidates for our needs. These species typically produce more milk than needed for their young, can be reliably bred and managed in human-controlled settings, and have relatively docile temperaments that allow for regular milking. Their digestive systems also enable them to convert plant matter humans do not eat, such as scraps (particularly in the case of nannies), into valuable, protein-rich milk, making them an especially important and practical asset to human communities.

As mentioned above, one of the reasons these species were chosen as milk producers for humans was their natural tendency to herd, which makes them amenable to group management. Besides all of the above, their ability to thrive in the various environmental conditions where humans settled—for instance, camels in arid regions and yaks in high-altitude areas—meant they provided crucial sustenance in environments where other animals may not have survived.

Sheep and goats, both alive and after death, have been and remain invaluable to traditional pastoral life and are among the most valued animals in the region. Sheep, descendants of the mouflon, are docile and flock oriented, while goats are curious, agile, and sociable. These creatures provided meat, milk, company, and other resources, such as wool, skin, and bones, which were repurposed for clothing, shelter, tools, and containers. For example, animal stomachs were used to hold milk, in some cases leading to accidental fermentation, a serendipitous discovery that transformed milk into tasty curds or cheese. Ewe's milk, although less abundant than goat's milk, is rich and fatty, making it ideal for fermentation. Nanny's milk, which is whiter in color, also yields high-quality dairy products. Up until recent times, if a mother had died or did not have milk and there were no wet nurses available, in some cultures, nanny's milk was given to babies—in some cases, directly suckled from the teats.[2]

As for camels, Bactrian camels, native to Eurasia and particularly the Gobi Desert, were another crucial resource for transport, food, and so on.[3] Their milk was often a vital source of sustenance in hostile environments. Camel cow's milk is thin and slightly salty; the animals' ability to drink salty water and consume sodium-rich plants gives their milk its unique flavor. In the desert, camel milk is a vital source of nutrition and hydration for herders, who collect only the amount needed for each meal, as the milk is best stored in the animal herself.

Moving on to the relationship between humans and horses, this is a great example of the development of a species from a predator-prey dynamic to a special companionship. Our earliest interactions were purely for consumption; we hunted these creatures for their meat, viewing them as a source of nourishment. The move toward domestication was a fundamental one—perhaps even a paradigm shift. It occurred when humans began to not only eat horses but also harness their physical attributes for transport.

Contemporary research has illuminated the amazing cognitive and emotional capabilities of horses, revealing them to be creatures of extraordinary awareness, impressive learning capacity, and notable memory. Their quick reflexes and innate navigational abilities, akin to an internal Global Positioning System (GPS), further distinguish them. Perhaps most intriguingly, horses exhibit sophisticated social intelligence, demonstrating an ability to interpret our emotions, which explains their increasing role in therapeutic settings through equine-assisted therapy. All these characteristics have made horses invaluable companions for the nomadic peoples of Central Asia then and now. They give their horses names in the same simple, descriptive style as dairy products, like "Patch" or "White and Brown." This importance of horses in Mongolian culture is particularly evident in the endangered Takhi, or wild Mongolian horses, which hold special significance for the Mongol people.[4] The word *Takhi* can mean either "holy" or "spirit," and these horses symbolize wild freedom.

In this region, horses transcended mere utility to become cultural symbols and measures of wealth. In the thirteenth and fourteenth centuries, the success of Mongol invasions and conquests culminating in the largest contiguous empire in history was because of their formidable cavalry, which proved instrumental in their dominance over much of Eurasia. In times of scarcity, they would even draw blood from their horses, diluting it with water or milk for consumption. They also consumed horse meat, considering it more reliable than other animal proteins. Their sophisticated knowledge of mare's milk processing led to the production of *kumis* (an alcoholic drink made with mare's milk), a topic that will be explored in greater detail in chapter 10. The historical relationship between horses and humans illustrates the potential for deep, multifaceted interspecies partnerships that shape both cultural practices and identity.

Based on our relationship with horses, it seems plausible that asses and donkeys and their hybrid offspring, mules and hinnies,[5] produced through horse-donkey crossbreeding, were also utilized. Female donkeys and asses are called jennies, and female mules, or mollies, may have been milked alongside other animals. However, I have found it hard to find evidence for this practice. On the other hand, the cultural significance of these animals, particularly in Āzarbāijān, manifests in fascinating linguistic and social dimensions, as well as in humor. In the Āzari language and culture, the word *eshayh*, or "donkey," serves as the primary pejorative in confrontational situations; if it gets very heated, you will hear *eshayhin bir eshayh*, which is "donkey of donkeys." This cultural embedding extends to regional stereotypes, particularly regarding the inhabitants of Tabriz. Tabrizis are often characterized as displaying mule-like obstinacy, a comparison that, as a self-identified Tabrizi, I and those who know me well can attest carries some truth in terms of determination and single-mindedness.

The humorous stories and anecdotes of the Turkic Mulla Nasruddin are famed in the region and often feature donkeys. Here is a typical example:

> A neighbour who Nasruddin didn't like very much came over to his compound one day. The neighbour asked Nasruddin if he could borrow his donkey. Nasruddin, not wanting to lend his donkey to the neighbour he didn't like told him, "I would love to loan you my donkey but only yesterday my brother came from the next town to use it to carry his wheat to the mill to be grounded. The donkey sadly is not here." The neighbour was disappointed. But he thanked Nasruddin and began to walk away. Just as he got a few steps away, Mullah Nasruddin's donkey, which was in the back of his compound all the time, let out a big bray. The neighbour turned to Nasruddin and said, "Mullah Sahib, I thought you told me that your donkey was not here." Mullah Nasruddin turned to the neighbour and said, "My friend, who are you going to believe? Me or the donkey?"

Other beasts, like buffalo, yaks, and aurochs, played a pivotal role in human history. According to some, they were initially used for traction before their value as producers of milk, meat, and hide was recognized. Aurochs, now extinct, were formidable oxlike creatures domesticated over ten thousand years ago and later bred into two main cattle types: Bos indicus, suited for tropical climates, and Bos taurus, adapted to temperate zones. Their milk is versatile and well suited for making yogurt, cheese, and butter. In my mother tongue, the term *ochuz*, today used for "ox," sounds similar to aurochs.

The buffalo, or *Bubalus bubalis*, is another significant contributor to human dairy history, particularly in temperate regions, where it has adapted well. These animals are thought to have been domesticated around eight thousand years ago. Known for their intelligence and fierceness, milking buffalo poses challenges, but the effort is worthwhile, as their milk is exceptionally rich and luxurious, with a high fat content. It is my favorite type of milk for making yogurt and *gaymākh* (thick cream).

Dri, like buffalo cows, are not the most approachable animals and are difficult to milk. Dri are protective of their milk and prefer to reserve it for their young, making the process of milking a significant effort and hard work. I enjoyed reading about how they are milked: Herders allow the calf to suckle briefly to stimulate milk letdown, then they sneak in to collect the milk. This milk is a vital source of sustenance at high altitudes.

Another factor influencing the health of all animals and, for the purposes of this discussion, mammals' milk, is the inherent ability to self-heal. You may have heard of people having cravings for odd things; more often than not, this is the body's way of telling us it needs a particular nutrient. We have also heard of babies eating soil or chalk. In the 1930s, in what would not be allowed today and deemed immoral, the scientist Clara Davis conducted an experiment on toddlers to study self-selection of their diet. Babies, some of whom had rickets, were offered cod liver oil, and those toddlers in particular ate it. Subsequent testing confirmed some had recovered from rickets!

In a similar way, animals can self-heal and bring their bodies into balance by seeking minerals and vitamins from their habitat. Animals who feed in natural settings are mostly browsers rather than grazers; they forage and eat only what they need. In different seasons, and in the case of nomadic herds who move to different locations or altitudes, the animals have a more varied diet. The milk from these types of animals is usually more nutritious and balanced.

Raw natural milk as described above is alive. It is a nutrient-rich liquid and a complex emulsion, primarily made up of water, fats, proteins, carbohydrates, vitamins, and minerals. This essential fluid provides young mammals with vital nutrition and hydration during their early development. Rich in both macro- and micronutrients, enzymes, and microorganisms, milk has often been regarded as having almost mystical qualities.

We would describe milk as an opaque white liquid; the characteristic color is a result of its fat and protein molecules scattering light. It is one of humanity's earliest and most versatile foods, also valued for its transformative properties, allowing us to make a variety of cheese, yogurt, cream, butter, and other dairy delights. It has been consumed in Eurasia for millennia, playing a critical role in spreading the genes, language, culture, indigenous knowledge, and beliefs of the people who relied on it. This was facilitated through trade, travel, and conquest.

In this book, when I refer to natural milk, I mean milk from heritage breeds of animals left to their own devices, raised in environments with little to no pollution, whether from soil, air, or water. These untouched habitats harbor indigenous bacteria, fungi, and yeasts that contribute to the nutrients, smell, and taste of this type of milk. Such milk varies daily, influenced by factors such as the flora and elevation of the environment as well as the health and hydration of the animals before milking. This milk, straight from well-kept animals without chemicals or heat processing, is a complete food and is safe to use. Natural milk has all its good bacteria, macro- and microorganisms, and enzymes intact. There is a balance of fats, sugars, and vitamins, rendering them more bioavailable for us. For example, it has vitamins K2 and D, which are best when taken in their most natural form and in combination. Similarly, the fat, sugar, and protein structures are easier to digest and keep you satiated for longer without affecting your metabolic health. On the other hand, natural milk can be a source of various bacteria, such as *E. coli*, *Salmonella*, and *Listeria*, which are dangerous for all but especially those with compromised immune systems.

The first time I ever had something comparable to natural milk was when we were visiting friends at their *bogh* (forest garden) in Iran. Their gardener, Mohabat, brought milk as a gift from his village near Līghvān. The sheep of this village browse on the slopes of Mount Sahand, and the cheese made from this milk is famous throughout Iran as being the best. Our hostess, Soraya Khanum, home pasteurized the milk, then brought it to us as elevenses (a midmorning snack). Not being a milk drinker, I initially declined, but when I heard the "oooo, ahhh" of the others drinking it, I had some. Wow! It was a revelation. The first thing I noticed was how sweet it was, so I asked if they had added sugar, but they had not. The warmth and fattiness of the milk allowed all the aromas and flavors of the flora of that hillside to be released in every mouthful. It was not just milk; it was a multilayered sensory experience. It is one of my memorable drinks, and to this day I feel it a great privilege to have tasted the place and maybe the past in that little glass of warm milk. Could it be that the milk, even though it was heated and had lost some of its nutrients, was appreciated by my body's bacteria, both from that part of the world?

In Central Asia, there are two primary words for milk that reflect the influence of dominant cultures throughout the history of the region. In areas shaped by Turkic influence, the word for milk is *sut*, while regions influenced by Iranian culture use *sheer*, from Fārsi, with different local pronunciations of *sheer*. Interestingly, the Turkic word for water is *su*. This makes me question whether the simplicity of words like *su* and *sut*, which almost sound like simple vocalizations, suggests that these are among the earliest words in these languages. Both substances, being essential to life, may have inspired similarly simple yet foundational terms. Similarly, I wonder if other simple Turkic words like *āt* (horse), *at* (meat), *it* (dog), and *ot* (fire) suggest the deep importance of these resources to early societies.

Beyond its linguistic usage, milk holds cultural, spiritual, and symbolic significance in Central Asia. Milk and its derivatives remain deeply respected in rituals, ceremonies, and social practices in many cultures. It represents spiritual abundance, nourishment, enlightenment, purity, and the interconnectedness of human life with the divine. Moreover, milk serves as a potent symbol of fertility, life, and cosmic renewal, deeply rooted in the shamanistic beliefs and spiritual understandings of the people of these lands. Much of this cultural and symbolic heritage is preserved through oral traditions rather than written records, with stories, myths, rituals, and customs informed by these beliefs.

Among the Mongols, milk has been integral to rituals and spiritual practices. Votive offerings of milk, such as sprinkling it into the air to honor the sky god Tengri or onto horses, are common.[6] In different parts of Eurasia, even to this day, milk is sprinkled on the bride and groom as a blessing in weddings, or it is shared from a single cup as a symbol of marital unity. Disrespecting milk, such as wasting it or stepping over it, is culturally taboo.

In these cultures, different types of milk are imbued with unique spiritual meanings. For example, mare's milk and *kumis* symbolize freedom, nobility, and the warrior spirit; ewe's milk represents community; and nanny's milk signifies adaptability and survival. These traditions are especially evident among the nomadic pastoralist descendants of Kazakhs, Uzbeks, Kyrgyz, and Mongolians, who include milk offerings and other delicacies in their Nowruz (New Year) celebrations.[7]

Not only the milk but the animals that provide milk are central to the lives of herding communities but also reflected in idols and gods within their belief systems. Some of these motifs have been adopted into the iconography of various pagan and universal religions. We have found cave drawings, figurines, and depictions of deities, some of whom were women with exaggerated breasts, which highlights the sacred status of life-giving milk. Milk and its derivatives remain deeply respected in rituals, ceremonies, and social practices in many cultures. In contemporary nomadic societies, individuals from dairy-producing backgrounds honor their traditions by maintaining a deep connection to their heritage, continuing to care for their special breeds and making dairy products in traditional ways, as seen among many nomadic tribes of Central Asia.

Even now *āghārty* are offered to guests as a gesture of respect and welcome, which is another way traditions are kept alive. This act also highlights the essential role of milk and dairy in the customs and social fabric of the region, where sharing these products continues to symbolize the dairying heritage. Needless to say, dairy also features prominently in culinary traditions, forming an essential part of the *dastarkhan* or *sufra*, which is a piece of material laid out for eating, like a tablecloth, with cultural significance. For example, my *sufra* is never without yogurt.

Milk and dairy products have also been an integral part of traditional medicine, often influenced by particular cultural beliefs or knowledge transfer among cultures. The milk of most but not all animals is regarded as a "cold" or "wet" substance in medicinal traditions across Eurasia. The schools of medicine are complex, and whether something is cold or hot often also depends on many factors, including the individual's body composition, temperament, health, climate, and diet. In simplistic terms, in Chinese medicine, milk is categorized as a *yin* food, offering cooling benefits. Similarly, in Ayurvedic medicine, it is recommended to balance *pitta*, and in Iranian Unani medicine, it is classified as *soyukhlukh* in Turkic languages, which means a food that cools the humors. As highlighted before, *āghārty* were mainly consumed during the summer in hot, arid climates, which may have contributed to their association with cooling properties or remedies.

Interestingly, milk and dairy's ability to cool or neutralize burning or heat can be put to use when consuming foods made with chili peppers. These can create a burning sensation in the mouth, and dairy, rather than water, can quell the discomfort. This is due to the protein and fat in milk, which counteract the capsaicin in chili, calming the receptors in the mouth that react to capsaicin and produce the sensation of burning.

Until recently, cow's milk and dairy products made from it were rare in Central Asia; today, most of the milk and dairy consumed globally comes from cows. What we now recognize as cows, which I call "modern cows," did not exist before; instead, their wild ancestors roamed the landscape. Over the centuries, selective breeding has transformed what were fierce and stubborn animals into the docile, high-milk-yielding cows we know today.

These modifications to the animals have consequently reduced their milk's flavor and quality. Generally, native breeds that are naturally adapted to their environments often produce superior milk, meat, and other by-products. Interventions in natural selection can disrupt these balances, and it often takes generations for livestock to adapt to such changes, resulting in, at best, temporary declines in quality. Moreover, these man-made cows are often artificially inseminated within three months of calving so that they can produce milk for much longer periods than is natural. Some of these animals, such as the Holstein cow (the most common type in the UK, a black-and-white cow with massive udders), have been created specifically to produce a large quantity of milk. Quantity, rather than quality of life, environment, or milk, is the primary motivator for such producers.

In researching this topic, I was struck by the number of studies on dairy herd psychology. Besides physical changes, temperament appears to be the other key factor in contemporary large-scale milk-producing animals. Today, cows, essentially genetically modified milk-production machines, calmly walk into industrial milking parlors to have their oversized, milk-laden udders emptied. This, combined with their ability to convert a wide range of what I would call unnatural fodder into milk, has made them indispensable to industrial dairy farming. The animals are fed a variety of natural and artificial products, which in most cases also include genetically modified crops. The animals are grazers; they can only eat what is put in front of them rather than foraging like the browsers of wild meadows presented above.

Advertisements for milk and dairy products present an idyllic image of cows grazing and naturally farting peacefully in open fields,[8] while the reality is often far removed from this portrayal. Terms like *grass-fed* appear on packaging, but their meaning is ambiguous and potentially misleading. For instance, *grass-fed* could refer to some dried grass being included in the feed. We do not know what percentage of the feed is grass versus other grains, such as corn or soy. We also do not know what else is in it, which could include synthetic and genetically modified elements, chemicals, and antibiotics. Not only that, but in wholesome advertisements, many companies offer consumers suggestions for how to enjoy the milk, usually alongside other packaged goods, such as cereals in the morning and cookies before bed. In most cases, this is done with little regard for the impact on our metabolic health.

In practice, much of the milk consumed globally is produced in large-scale dairy barns where cows are fed genetically modified grains, subjected to chemical treatments to prevent disease, and kept in sanitized environments designed for maximum productivity and minimum variation. Many of these animals never experience sunlight or graze on fresh grass, instead spending their lives in controlled settings to ensure "happy, high-yield" milk production. Tragically, the industrialization of dairy farming comes with risks, as evidenced by the catastrophic explosion at a Texas dairy in 2023, which killed *eighteen thousand* cows:

> Investigators believed the fire might have started with a machine referred to as a "honey badger," which [Sheriff Sal Rivera] described as [a] "vacuum that sucks the manure and water out."
>
> [Rivera continued,] "Possibly [it] got overheated and probably the methane and things like that ignited and spread out and exploded."

Such words could have been said by Chief Wiggum from *The Simpsons*!

Industrial milk production addresses hygiene through standardization and sterilization, creating a uniform product with the efficiency of fast-food operations. Pasteurization, developed in the mid-nineteenth century by French chemist Louis Pasteur, revolutionized food safety. Originally devised for wine, it gently heats liquids to destroy harmful microbes without compromising flavor. In 1909, industrialist J. D. Rockefeller lobbied for pasteurization to be applied to milk, in the process restricting sales to producers with pasteurization facilities, generating fear about raw milk, and limiting public access to it. Modern methods heat milk to specific temperatures for defined durations, such as 63°C (145°F) for thirty minutes or 72°C (162°F) for fifteen seconds.

There is also ultra-heat-treated (UHT) milk, which is sterilized completely. In this process milk is heated to 135°C–150°C (275°F–302°F) for a few seconds, which kills all bacteria and some bacterial spores. Subsequently, this type of milk can be stored at room temperature for months. Heat processing milk strips away many of the nutrients—for example, vitamin D—some of which are then added back in a process known as fortification. These so-called advancements often destroy beneficial microorganisms as well as harmful ones, much like the broad-spectrum effects of antibiotics.

There are a host of other modern methods of making milk safe for consumption, which include using ultrahigh-pressure treatment, pulsed electric field technology, sound waves, biocontrols such as viruses against harmful bacteria, and even essential oils and gentle water bubble motions, to name a few.[9] Producing safe natural milk requires knowledge, care, effort, and time, all of which are bypassed by these modern industrial processes. Industrially produced raw milk as a whole is not suitable for consumption straight from the animal; it has to be heat-treated to render it usable.

Before modern, industrial, safe milk was available, it was either boiled or scalded to ensure safety, a practice still followed in my home in Tabriz. Most families still purchase raw milk from a local dairy shop by weight. This is milk that hasn't been heat-treated but may have had some of the fat removed. It is not whole milk, and sadly it isn't natural milk either, as the animals may have been treated with pharmaceuticals and fed industrially produced fodder as well as some grass rather than natural pasture. The way we pasteurize milk at home is to heat the milk to 63°C (145°F) and keep it at that temperature for twenty to thirty minutes.

Where raw milk is readily available, the tradition of careful milk handling resembling the home pasteurization outlined above persists in many countries. In Turkic languages, we call it *pishirmakh*, which means "cooking." You see the term *scalding* in vintage English cookbooks, reflecting the common practice of heat-treating milk and sometimes equipment to sanitize or improve food texture.

Usually this type of raw milk has had some but not all the fat removed, and it has not been homogenized; therefore, when this heated milk cools, a skin known as lactoderm forms on the surface, which is both a nostalgic and divisive feature in our home.[10] My father loved it and, with great glee, scooped it up with a sugar cube.

Homogenization, as just mentioned, is a process where milk fat particles are altered to create a uniform product so that the fat in milk does not separate. On that note, most modern milk is available in four levels of fat: whole milk, with around 3.25 percent milk fat; reduced fat, with 2 percent;

Home pasteurizing milk: lactoderm puffing up

low fat, with 1 percent; and fat-free or skimmed. To create these types of milk, first, centrifugal force is used to separate the fat, which is then added back in at specific percentages for each type of milk, after which it is pasteurized, homogenized, and packaged.

During this research, I came across two other distinctions for mass-produced milk that are not talked about widely: A1 and A2 milk. These are distinguished by the type of beta-casein protein they contain. A1 milk, common in modern cattle breeds like Holsteins, contains both A1 and A2 beta-casein proteins. In contrast, A2 milk contains only the A2 form, a more naturally occurring protein in milk that is also found in human, nanny goat, and ewe's milk and comes from breeds such as Jersey and Guernsey cows. Historically, the A2 protein has been the standard in cow's milk for thousands of years, but a genetic mutation in modern breeds such as the Holstein and Friesian led to the emergence of the A1 protein. A2 milk, being more natural, is easier to digest. There is even an A2 Milk Company Limited in New Zealand that commercializes A1 protein-free milk and related products, including infant formula, under the A2 and A2 Milk brands.

Buying milk in Central Asia, and Iran in particular, is a little bit different from buying it in the West. There is still a distinctive commercial style characterized by small, specialized independent retailers, an economic model that contrasts sharply with the consolidated, impersonal supermarkets prevalent today. These shops create a network of expertise, with vendors focusing exclusively on specific categories like poultry and eggs or dairy. There are a few large supermarkets, but overall, in Tabriz at least, the majority of retailers are independently owned shops.

I introduced you to Mr. Abdi in the introduction to this book; we buy almost all our dairy from him, whether it is milk, butter, yogurt, or cheese. We know each other, and there is a relationship. He knows what we like, so if he has a particularly good batch of buffalo cow yogurt, when I am in Tabriz, he drops it off at our house or texts my mother to let her know. The products he stocks are artisanal and are not branded; they are known as *achikh*, or "open/unpacked." He has large blocks of cheese and butter and refrigerated barrels of milk. He weighs out the amount of raw milk we would like to buy into a plastic bag that we bring home to scald. To end this section on rendering milk safe, the most peculiar way to do this is with an amphibian. Apparently, some frogs have peptides on their skin that keep milk fresh!

Mr. Abdi in his dairy shop

Besides drinking milk and making dairy products, there are some other applications too. I cannot vouch for any of these, as I have not tried most of them, but they are the most popular I came across: thawing frozen fish in it to retain its flavor, boiling corn in it to enhance sweetness, and using it in skincare practices, such as in Cleopatra's legendary milk baths. Milk can polish leather, nourish houseplants, and be added to compost for nutrient enrichment. Its protein, casein, is used in manufacturing plastic (unsurprisingly called "milk plastic") and as a glue in repairing broken porcelain, to name a few uses. I remember when, as children, if we were stung or had sunburn, my mother would apply yogurt, which had a cooling effect, and I must say it worked.

When we think about it, it is quite remarkable how milk is available and consumed all over the world now. In Southeast Asia, the Americas, and Australia, milk and dairy products were historically not a large part of the diet of the indigenous population, for a variety of reasons. Milk consumption became common in these regions comparatively recently, mainly as a consequence of European colonialism over much of the world in the last five hundred years. European diets, government interventions, economic policies, and legislation have all played crucial roles in shaping dairy-consumption patterns. Understanding these political factors provides insight into the current global production and distribution of dairy and dairy products.

Now India is the top producer of all types of dairy, while the United States is the highest manufacturer of cow's milk, and China is the main producer of ewe's milk. In recent decades, the dairy industry has faced increasing scrutiny due to concerns about animal welfare, environmental impact, and public health. As food systems continue to transform, it is crucial to address the ethical, environmental, and social dimensions of these models.

Plant-based milk, sometimes called "mylk," has grown increasingly popular in the West in recent years, fueled by food trends and a backlash against dairy. There is a similar increase in the variety and availability of nondairy "cheese," "butter," and "cream." These are, of course, also made for people choosing plant-based diets and for those individuals who cannot consume dairy. Historically, simple plant-based milks made from soy and almonds have been used in China and Central Asia. Modern versions are derived from a variety of nuts, grains, and beans, and some plant-based milks are also synthesized by fermenting dairy whey and combining it with plant-based ingredients to mimic the qualities of traditional dairy milk. However, commercial versions are often highly processed and contain numerous additives.

Plant-based milk can be easily made at home with minimal equipment and raw materials, using nuts or seeds and nonchlorinated water. For instance, blending 150 grams (5 oz.) of raw nuts (soaking them beforehand is always a good idea) with 1 liter (2 pt.) of water and straining the mixture through a cloth-lined sieve produces a nut milk. The residue left behind can be turned into a sweet or savory nut spread with additional flavorings. The history of milk production illustrates human ingenuity shaped by necessity and cultural adaptation over centuries. An examination of its development reveals stories of survival, innovation, and evolution, offering valuable insights that can be applied to contemporary challenges and aspirations. The stark contrast between romanticized imagery and industrial reality underscores the ethical and environmental dilemmas surrounding modern dairy production. Reflecting on this trajectory, it is ironic that while humanity has achieved extraordinary technological feats, such as landing on the moon and developing artificial intelligence, we have simultaneously degraded the quality of most of our food and ecosystems.

From the moment humans began exerting control over nature, we have consistently depleted the land and denatured the animals and products we consume for short-term gains. This has often been done with little regard for the well-being of the land, animals, or even ourselves—all in the name of progress and growth (and I would go so far as to say, nowadays, greed or laziness as well). In the UK alone, half a billion tons of milk are wasted annually,[11] and this is just one small country. Scale this waste globally, and the picture becomes even more staggering.

Setting aside, for a moment, the ecocide and ethical outrage of wasting food in a hungry world, consider the long journey of wasted milk and dairy products. It begins with a cow, engineered, monitored, and milked; this is followed by testing, analysis, packaging, refrigeration, transportation, and placement of the milk on store shelves. From there, it is purchased (often by someone who drove to the shop), brought home, refrigerated again, partially consumed, and then discarded along with the packaging. Imagine the resources and energy expended throughout this entire process. And yet this is only part of the picture. The downstream effects of wasted dairy on the environment, where it ends up, and how it impacts ecosystems are other troubling factors I have not even begun to unpack.

Viewed in this way, the entire system starts to feel like an elaborate con. In a world where fridges communicate with phones and devices can interpret our thoughts, why not leverage technology or AI to predict the precise amount of milk required by local populations worldwide? This data could be relayed to local farmers, enabling them to produce milk on demand for either local pickup or delivery. Customers could bring their own containers or have milk delivered in reusable bottles.

By eliminating milk waste and producing only what is needed, we could allocate the appropriate amount of land and resources for dairy production. This approach would ensure everyone who desires it has access to fresh, traceable, nutritious milk.

Having said this, I am somewhat encouraged when I read that not only are many dairy farmers around the world trying to get back to more traditional methods to care for the land, animals, and ultimately our health, but also some are trying to breed cattle that resemble old breeds.

I will conclude this chapter by sharing an embarrassing personal story relating to UHT milk. I do not know if they still make long-life milk in small, soft, plastic cups sealed with a foil top anymore, but there was a knack to opening them safely. If you squeezed them too hard when trying to pull the foil off, the milk would squirt out and make a mess. My mother and I were having coffee in the late 1980s in London, and a gentleman at the table next to us was struggling to open the awkward little packet. As we watched, I said to my mother, in Āzari, that I could not understand why people struggled with these packs and that there was a little piece you just had to pull back. To my embarrassment, the gentleman riposted in Āzari, "I have arthritis in my hands, which makes it hard for me, my daughter." Talk about wanting the ground to open up and swallow you! My mother and I both turned bright red, and I apologized profusely for my inconsiderate remark. This experience serves as a compelling illustration of several sociological principles: the danger of making assumptions about others' capabilities, the false sense of privacy that speaking in a less-common language can provide, and the way that casual ableism can manifest in everyday observations. Since then, I have tried to be more considerate and careful.

3

Fermentation, a Microbial Marvel

Thus far we have explored our evolutionary journey to herding animals and consuming their milk. We have seen how some believe that it was fermented foods, not fire and cooked foods, that accelerated our development—that is, our physiology, cognitive abilities, and behavior. Fermented foods have been integral to human diets since the earliest days of our existence, including both plant- and animal-based foods. In this chapter, we examine food fermentation, why and how it happens, with a particular focus on dairy fermentation. To do this, we will delve into milk's natural propensity for transformation, examine the components of milk, and explore how bacteria, yeast, and enzymes interact with milk to transform it into products like yogurt, cheese, and kefir. I will also talk about some of the health benefits and drawbacks of fermented dairy foods. This chapter ends with practical advice and basic requirements for home fermentation.

The intricate interplay between microorganisms and milk produces a variety of fermented dairy products, many of which hold significant cultural and culinary importance to this day in Central Asia. Fermentation, while potentially risky, offered our ancestors a relatively predictable means of transforming food to make it safer to eat and make it last a bit longer. Unlike hunting or gathering, fermentation required no other action once set into motion, allowing humans to focus on other survival tasks. Over millennia, we developed the intellectual capacity and technological means to deliberately ferment food.

While the exact origins of deliberate fermentation remain unknown, the archaeological evidence we have to date suggests that humans have been fermenting food and drink, including fruits, vegetables, grains, legumes, dairy, fish, and shellfish, for at least nine thousand years. Fermentation is practiced across diverse climates, from tropical regions to arctic environments, and utilizes a wide range of naturally occurring microorganisms. It happens over a range of time, from a few minutes to years. Depending on the timescale, fermentation can enhance food flavor in the short term or serve as a long-term preservation method. It also increases the safety and nutritional profile of foods—for example, certain plant compounds that cause irritation if ingested (lectins, which are part of a plant's defense system against herbivores and parasites, are weakened by the process of fermentation).

In the past, one of our most precious items was our funky ferments. I am not just using the word *funky* to alliterate; rather, most ferments do have a certain aged aroma and tangy flavor. Just like seeds, they were future food, something to respect, look after, eat, share, and pass down through the generations. If we think about it, it is quite marvelous that a "starter" like kefir grains keeps multiplying and can be easily shared. No wonder fermentation starters are also called the "mother." Long before we knew the actual microorganisms involved in fermentation, it must have seemed like magic. You would have left milk overnight and returned to find it had turned into curds and whey and that it smelled and tasted different.

The pastoral herders of Central Asia have always relied heavily on their flocks, with milk often serving as their sole source of sustenance in harsh weather or hard times. Human consumption of fermented milk likely began shortly after we started keeping ewes and nannies. As mentioned before, milk itself naturally ferments due to microorganisms present within it and those in the surrounding environment, which could originate from the milker's hands, containers, plants, insects, or airborne yeasts. Each instance of fermentation produces a unique ferment regardless of the substrate, like a photograph capturing the time and place. The evolution of fermented milk products has been shaped by the materials, climate, and technological ingenuity of our forebears.

Our ancestors navigated a world vastly different from the one we inhabit today, and by examining one of their key sources of nourishment, dairy, we can trace the origins of fermented dairy products and their evolution into the butter, kefir, and other foods we still enjoy today. They developed practices through trial and error, which were passed down and have become indigenous wisdom. Resources such as fermentation starters and knowledge were shared, which helped them and their progeny survive and endure harsh conditions. For our omnivorous ancestors, fermentation provided a vital means of adapting to environmental changes, preserving food, and increasing survival odds. As their knowledge expanded, fermentation techniques were applied to a broader range of products, including medicines, animal skins, and clothing.

We gradually created different methods of producing and preserving dairy products, largely through accidental discoveries. When ceramics appeared in western Eurasia, people used them for milk. Pottery advances coevolved with dairy processing, enabling innovations such as beeswax-coated strainers. These developments allowed for more effective management of curds and whey, leading to easier dairy processing.

When humans migrated, they took their ferments with them. Whether it was dried chunks of dairy or pieces of material soaked in fermented milk, they transported, shared, and kept the ferment going. For example, they soaked rags in yogurt and then dried them. When they eventually settled, they put the rags in warmed milk and started the fermentation process so that the ferments were resurrected when needed. The microorganisms within the ferments responded differently to the new environment and, in some cases, further evolved.

Dairy ferments became more stable and robust as time passed. One of the ways in which this could happen was by reusing vessels like gourds to contain milk. This allowed a mix of microorganisms to colonize the inner surface of the gourd, resulting in more consistent and reliable ferments.[1] The ferments were carefully tended, much like their animals, which provided milk.

Fermentation techniques and the ferments themselves spread and were adapted to local ingredients and tastes. Every country has many local ferments; some are part of the national identity, such as tempeh from Indonesia or *kumis* from Kazakhstan. Today, the rapid exchange of information through social media has accelerated the spread and use of ferments around the world, giving rise to fusion dishes like kimchi pasta.

Fermented foods are also known as cultured or live foods, terms rooted in the Latin *cultura*, meaning "to care for" or "cultivate." Over centuries, the term *culture* evolved to encompass microorganisms and, later, human achievements and behaviors. *Live* refers to the active microorganisms within these foods, which need to be looked after to remain alive.

The science of fermentation, known as zymology or zymurgy, originates from the Greek word *zym*, meaning "fermentation." Some of the most popular foods around the world, such as chocolate, olives, coffee, tea, wine, beer, yogurt, and cheese, are time-honored ferments.

The English word *fermentation* derives from the Latin *fervere*, meaning "to boil, leaven, or rise." This aptly describes the visible activity during fermentation, such as bubbling, gas formation, and effervescence. In Fārsi, fermentation is *takhmir*, an Arabic word. In most Turkic and Iranian languages, there is no singular term for fermentation; instead, people describe the outcome of each ferment. For example, yogurt, cheese, and vinegar are said to have "arrived" (*jalip*); curd is described as having been "caught" (*tutup*); and sour milk "curdles" (*churyip*). Similarly, in Iranian languages, they describe yogurt as having "closed up or set" (*mast baste*), with *bastani* meaning "a firm or closed thing," or ice cream.

Fermentation is a fascinating biochemical process, and at its core, it occurs when microorganisms break down compounds, often carbohydrates such as sugars, under particular environmental conditions, typically in the absence of oxygen. The bubbling, leavening, and boiling often associated with fermentation arise from the release of energy during these chemical reactions, producing by-products such as alcohol, acids, and gases. For example, bacteria transform lactose into lactic acid, imparting the characteristic sourness found in yogurt or sourdough bread. This transformation gives the final fermented product a distinct appearance, taste, and texture. We could say fermentation partially predigests food by breaking down complex compounds into simpler, more bioavailable forms, which can make them easier for the body to process.

Fermentation embodies life and transformation, and to me, it is akin to gardening. In our organic garden and allotment, plants often grow without being intentionally planted, carried there by the wind or animals. When they land on fertile soil, they thrive. When such plants grow in undesired spots, however, we call them "weeds." To cultivate healthy plants, you need good soil, water, viable seeds, and optimal conditions; with luck, and only at the plants' natural time, they bloom or ripen, allowing us to harvest. Each year, these conditions and timings vary slightly. For novice gardeners, the process can be fretful, but with time, they gain confidence and knowledge and start to trust their instincts. I believe fermenting follows a similar journey. Just as some individuals delight in growing fruit, others prefer flowers; likewise, in fermentation, some enjoy baking sourdough, and others prefer making cheese.

Fermentation can occur in two main ways: wild or spontaneous fermentation and controlled fermentation. Wild or spontaneous fermentation is an unpredictable, natural process that happens without the need for intervention. Controlled fermentation, by contrast, involves deliberate action and the use of specific microorganisms and conditions to guide the process. Fermentation begins when invisible microorganisms come into contact with raw materials or when a starter culture is introduced.

We have seen that milk has a natural propensity to ferment; that is, if we leave fresh milk undisturbed, the microorganisms present in the milk, along with the temperature of the room and naturally occurring bacteria and yeasts in both the milk and the environment, will eventually cause it to curdle, and this is known as clabber. The word *clabber* means "to become like mud." I quite like Harold McGee's "fragile solid": To me, this is a great description. Clabber is not very flavorsome and gets more sour as time passes.

Despite being equipped with sensitive olfactory systems, our ancestors learned the hard way what to eat and what to avoid. There must have been many deaths from ingesting or being exposed to pathogens in food, air, water, and milk. They slowly learned to distinguish between good and bad food, including clabber and spoiled milk.

In the gardening analogy above, spontaneous or wild fermentation is akin to a "weed" taking root in a garden by chance. On the other hand, controlled fermentation is more akin to purposeful gardening, like choosing to grow roses or to make kefir.

Today, if milk alters in form, whether it is scent or taste, such changes are often viewed with suspicion, and it is discarded for fear of illness. However, our ancestors were far less wasteful. Having consumed the altered milk and survived, they would have incorporated it into their diet. Clabber is the first stage in milk's transformative journey.

Milk is mainly water with macro- and micronutrients, and water is a breeding ground for bacteria. Freezing, drying, adding salt or sugar, or making the environment acidic makes free water an inhospitable environment for pathogens and generally resistant to spoilage and toxins.

Prior to modern preservation methods, such as canning, jamming, and freezing, fermentation enabled us to cache food and drink for another day or for hard times. It was nourishment if the weather suddenly changed, if we fell ill, or if we were far from sources of food and drink.

Fermentation is a simple yet highly effective method of food production, transforming basic raw materials, such as fruits, grains, or animal proteins, into functional, flavorful, and more nutritious foods through microbial and enzymatic activity. There are two primary types of lactic acid fermentation. First, when lactic acid is the only fermentation outcome, the process is said to be homolactic fermentation, as is the case with yogurt. Second, with heterolactic fermentation, besides lactic acid, there are also other by-products, such as ethanol and carbon dioxide, as in kefir and both types in sauerkraut.

Modern science has identified and categorized many of the strains of bacteria involved in dairy fermentation, such as *Streptococcus thermophilus* and *Lactobacillus bulgaricus*, which are industrially utilized to produce standardized products like yogurt. These microbes can be used alone or in various combinations to create a vast array of fermented foods. The most common microorganisms involved in food fermentation include lactic acid bacteria (LAB), lactic acid bacteria, yeasts, and fungi, either singly or in combination.

For milk to curdle, sour, or clabber, sugar-loving LAB, which are present in milk itself, land in milk, or are on the surface of a container into which milk is poured, convert the milk sugar (lactose) into lactic acid. The resulting acidulation causes the protein in milk, casein, to separate into curds and whey, changing the chemistry, texture, and flavor of milk. This acidic environment, with lower pH, also inhibits the growth of other harmful pathogens, making the ferment last longer, the nutrients more bioavailable, and the product safer to consume.

Another set of microorganisms necessary for fermentation are yeast. These are single-celled, microscopic fungi that also play a crucial role in transforming milk into a variety of delicious and nutritious fermented dairy products. They thrive in all sorts of different environments and are versatile fermenters. They are essentially nature's alchemists, converting milk sugars into a range of flavor compounds and transforming liquid milk into products with unique textures and tastes. Yeasts work alone or alongside LAB to break down lactose and produce various by-products by consuming milk lactose and converting it into ethanol, carbon dioxide, and other metabolic compounds. *Kumis* is an excellent example of yeast-driven fermentation, where the yeast converts sweet mare's milk into a mildly alcoholic drink.

A famous enzyme in dairy fermentation is rennet, specifically used in cheese making, which gives us the largest variety of a fermented dairy category that is cheese. Rennet is a complex set of enzymes primarily derived from the fourth compartment (abomasum) of the stomach of young ruminant animals, particularly unweaned lambs, though now calves' rennet is also made. The key enzyme in rennet is chymosin, which coagulates milk proteins to produce curds and whey. It is said to have been discovered when milk was stored in containers made from the stomachs of lambs or kids. Basically, rennet works by targeting casein in milk. Chymosin cleaves kappa-casein, which causes milk proteins to alter and form more stable curds. This process is fundamental to cheese production.

Nowadays, there is microbial rennet synthesized through the fermentation of specific fungi or bacteria. These genetically modified microorganisms are used in industrial cheese production because they are cheaper and more reliable for mass-producing standardized products. There is also genetically engineered rennet, which is created by inserting calf rennet genes into microorganisms. This process makes a chymosin-like animal rennet and is said to address ethical and supply concerns regarding traditional animal rennet.

There is also a bewildering array of milk coagulants, such as ant eggs, ash, fig sap, nettles, the calyx of certain capsicum, lemon juice, vinegar, whey, and kefir, among other things, that are used to create curds and whey. As I said before, to reduce confusion in terminology and to maintain terminological consistency, I've decided to use the word cheese/*panir* only when talking about curds made with rennet. This is to distinguish it from all other methods of creating curds, however much they may resemble cheese.

I have touched on this before, but there is also mixed fermentation, in which several different microorganisms, such as yeasts and bacteria, work together to produce a ferment. These are also known as a SCOBY (symbiotic culture of bacteria and yeast). In this type of fermentation, yeasts convert sugar into ethanol and carbon dioxide, making the ferments fizzy, while the bacteria convert lactose into lactic acid, giving it its sour taste. Mixed fermentation is said to be the most beneficial for our health. Kombucha (a fermented tea) "mother" and kefir grains are examples of a SCOBY; these starters are tangible, visible, and reusable.

Here, I will briefly talk about another fermentation starter that looks, feels, and acts like kefir—that is, water kefir, known as *tibicos* or Japanese or Tibetan crystals. It, too, is a SCOBY just like milk kefir grains, is granular, is more translucent than milk kefir grains, and multiplies. Water kefir has different strains of yeasts and *Lactobacillus* bacteria than the ones in dairy kefir. The probiotic composition of water kefir differs from that of milk kefir grains. Consuming both provides a broader diversity of beneficial microbes.

The substrate or food for the water kefir necessary for it to survive is water and carbohydrates, which can be sugar, honey, fruit juice, or plant-based milk (with added sugar). Water kefir will not survive in or ferment dairy milk, whereas milk kefir grains can ferment plant-based milk. On average you need two tablespoons of water kefir grains and sugar or sweetener to a liter (2 pt.) of liquid. Sweetened coconut milk is an excellent substrate for water kefir.

For successful fermentation, the right environmental conditions are essential: Temperature, oxygen availability, duration, and light levels all play crucial roles. As an example, we know that if you heat most dairy products, including milk, some of the bacteria die, particularly the heat-sensitive ones, rendering them less nutritious. Therefore, we must be careful when applying heat, for example, to dry *gooroot*. When drying it—be it in the open air, indoors, or in a dehydrator or an oven—provided the temperature stays below 45°C (113°F), it will keep its bacteria intact. Similarly, cooking with yogurt can reduce some of its probiotic content, since the temperatures involved usually exceed 45°C (113°F).

LAB can ferment milk in both aerobic and anaerobic environments. Aerobic (with oxygen) fermentation is quicker, while anaerobic (without oxygen) fermentation takes longer. Fermentation microorganisms are also classified as either mesophilic, preferring temperatures between 20°C and 45°C (68°F–113°F), like kefir, or thermophilic, which thrive at 45°C (113°F) and above, such as yogurt. Most fermentations take place in cool, dark places over time.

In traditional practices, a starter is introduced to milk when it reaches "finger-biting" temperature, or *barmakh dishlian* as we say in Āzari, which is around 45°C (113°F), meaning hot enough that you quickly withdraw your finger without burning it. In suboptimal conditions, the fermenting microorganisms may slow down, become dormant, or die altogether. It is another reminder that these processes involve living entities.

Now let us turn our attention to dairy fermentation and milk's properties as a suitable substrate, starting with its three primary macronutrients—fat, protein, and sugar, which are necessary for fermentation. Good fats are necessary for our well-being, and some of our most important organs are made up of fat, such as our brain. Milk fat is mostly made up of triglycerides, cholesterol, and fat-soluble vitamins, among other elements. Fat's unique structure plays a big role in fermentation and influences the final flavor, texture, and aroma of dairy products. Some of the fermentation bacteria use enzymes to interact with milk fat, breaking it down into fatty acids that create flavor compounds and affect texture. Even the membranes around fat globules have a job: They provide a surface on which bacteria can grow, are necessary for the ripening process of aged cheese, and even affect the color of some dairy products. For example, the abundant fat in buffalo milk gives it a creamy texture and enhances the complexity of flavors.

The next macronutrient we will look at is protein; it is an essential nutrient for our health and comes in animal and plant forms. Protein is made up of amino acid chains that our bodies use to build and repair muscles and bones, and it is essential for a host of metabolic and endocrine functions. It is a source of energy and makes us feel fuller for longer. Any extra protein we digest is first converted into sugar and then stored as fat. Casein, the protein in milk, keeps the oils, fats, and water emulsified in milk and is responsive to changes in the pH balance or acidity in the milk. The remaining 20 percent of milk protein is whey. For example, when making yogurt, the casein acts as a gel-forming protein, leading to a more viscous product, and provides structural integrity. Casein in both curds and whey also contributes to flavor complexity, enhances nutritional value, and supports bacterial growth.

As previously mentioned, lactose, or milk sugar, is the primary carbohydrate in milk and a key energy source for microorganisms during fermentation. Lactose is a disaccharide (made of glucose and galactose) unique to mammalian milk. Moreover, bacteria and yeast (either wild or in starter cultures) break lactose down and, in the process, create an acidic environment. Mare's milk has the highest lactose content, and it is not usually consumed raw, as it would cause great discomfort; instead, it is fermented first.

Unsurprisingly in modern mass-produced milk, these three macronutrients—that is, the levels of sugar, fat, and protein in milk—as well as the rest of the cow's milk are closely monitored and maintained at pre-set levels.

To end this section on macronutrients in milk, they are now isolated and used in a range of industrial applications, such as in the manufacture of protein plastic—for example, paint, glue, and nanotechnology components. The whey is used to add protein to all sorts of animal and human foods as well as in the production of supplements.

Having looked at milk in general, now we will briefly examine each animal milk type, focusing on its suitability for fermentation, beginning with the most common milk-producing animals of Central Asia. It is interesting that, in general, these languages do not have distinct terms for male and female animals. Instead, they use gender descriptors like *archayh* in Turkic and *nar* in Iranian for males or *dishi* in Turkic and *māde* in Iranian for females. However, there are two notable exceptions. For instance, the Āzari term for a ewe is *goyun* and *goosfand* in Iranian languages; in both languages, these are the default or common terms used to refer to these animals. This may be due to the predominance of ewes in flocks or their greater cultural and economic significance in these societies. There is also the word for a ram or male sheep, *goch*, which is the same in both languages but is not used very much. The second exception pertains to mares. While the specific term for mare is *maydan* in both languages, the more commonly used word for horse is *āt* in Āzari and *asb* in Fārsi. I've often wondered whether the English word *maiden* might share etymological roots with *maydan*.

Before I talk about the milk of animals, I have to declare my fondness for sheep and goats. If you have ever looked a goat in the eye, you will notice their captivating rectangular pupils, or if you have walked behind a fat-tailed sheep, you may have been mesmerized by the pendulous movement of its tail and thought, like me, of Sir Mix-a-Lot's lyrics: "I like big butts an' I cannot lie."

When we moved to Wales, where you see many sheep on hillsides, these creatures were different in form and color and stood out against their lush, green, hilly backdrop, unlike what we were used to seeing. We had come from a place where sheep were always in droves, with a shepherd and sheepdog nearby, browsing for things to eat on mountains that looked lifeless from afar but up close were covered in a variety of flora.

Another fond sheep-related memory was on a visit to Lighvān, where we got caught in what I call "sheep rush hour": The shepherds had come back from the hills in the evening and were guiding sheep back into the village. As they entered the village, each sheep went to its owner's home. There is a British saying, "following like sheep." Maybe this is true when out in the mountains, but it didn't seem to apply when they were returning home.

Now let us examine each milk by animal; the words in brackets are the female animal name in Turkic and Iranian.

- Ewe's milk (*goy/goyun* & *goosfand*). This milk is prized for its creamy white color, high protein concentration, and rich fat content, making it highly compatible with fermentation. It is good for creating dense, flavorful dairy products.
- Nanny's milk (*dishi jechi* & *boze māde*). This milk is white, with smaller fat globules and a lower protein content. Its finer nutrient structures allow for faster fermentation, though it produces a product with a thinner consistency and a distinctive aroma.
- Buffalo cow's milk (*dishi jāmush* & *govemishe māde*). Often regarded as the most luxurious milk, it boasts the highest fat content, nearly double that of others, alongside superior protein concentration and lower cholesterol. Its dense molecular structure is good for fermentation, resulting in robust, stable curds.
- Mare's milk (*mādyan*). This milk is thin, high in sugar, and has the lowest fat content of these milks. Its complex sugar composition makes it ideal for fermenting into *kumis*.
- Camel cow's milk (*dishi dava* & *shotoreh māde*). This is a slightly salty, frothy, thin milk. Its unique protein structure and low fat content make it challenging to ferment, though it is used to make *shubat*, a drink similar to *kumis*. Researchers are exploring ways to identify the most suitable microorganisms for fermenting camel cow's milk.
- Dri's milk (*ghazh gāv*).[2] This milk has less lactose and more protein and fat than cow's milk and yields similar products to ewe's milk.
- Cow's milk (*inayh* & *gāv*). This is now the most commonly used milk in the world, and thanks to its manufactured balanced composition of protein, fat, and lactose, it is a good substrate for fermentation and forms the basis of numerous fermented dairy products.

I want to talk briefly about plant-based "milk" as well. I know our notions, norms, and language evolve over time, but around 2019, if someone said "milk," we'd think of mammal juice; now, in the West, we might also think of plant-based milk as well. It has been a very quick change. While this shift to plant-based milk is often marketed as eco-friendly, it is worth noting the environmental, economic, and human costs of some plant-based ingredients. Most, like cashews or soy, are often imported to the UK; apart from the most talked-about transportation costs, they additionally deplete groundwater required to grow them. Moreover, when shelling cashew nuts, workers' skin comes into contact with the caustic oils present on the hard outer shell, which can cause severe burns and blisters on their hands.

As a gardener and cook, I value eating locally and seasonally, and I encourage my students to do the same. When I started Simi's Kitchen, my cookery school, it was partly inspired by a friend whose children were diagnosed with celiac disease and were also lactose intolerant. Since then, I have become more aware of food allergies and intolerances, and I like to experiment to create dishes suitable for all. I have always aimed to offer alternatives that mimic the original dishes as closely as possible, but I have given them a different name. I respect plant-based cooks for their creativity and adaptability, even if I sometimes find it annoying when plant-based products are lazily named after their meat or dairy counterparts.

To conclude this section, plant-based milks, while great for those with allergies, aren't nutritionally equivalent to dairy. Most are made by soaking, grinding, and straining nuts, seeds, or grains and often include additives like oils and stabilizers. Even pure options, like some local UK-based fresh oat milk, can have a high glycemic index, which is not ideal for long-term health. Plant-based milks can be fermented, though the results differ in texture and flavor compared to dairy. The resulting curds and whey have some of the health benefits but lack the diversity of microbes, vitamins, minerals, and protein found in dairy.

I'm reminded of the joke about supermarkets having a "health food" section, which raises questions about the quality of the rest of the store's offerings. The less said the better. Nowadays most people live in cities away from farms and orchards, and many city dwellers are unaware of what foods are in season locally. Alarmingly, many urban children are unable to identify basic fruits and vegetables. Here in the UK, most food is bought in supermarkets rather than, say, from a greengrocer, though some supermarkets highlight seasonal produce. Although we have greater access to a wide variety of food, many people in the West are overweight yet malnourished.

Over the years, there have been many different types of trendy health foods. At the moment, the focus is on fungi and fermented foods; we are told they offer significant benefits for our overall well-being. Much like seeds in a garden, these foods must be received by an environment capable of accessing and optimizing their value. This underscores the importance of maintaining a healthy body with good microbiota, as these microbial communities play a critical role in accessing nutrients in the foods we eat. Luckily, fermented foods not only provide nutrition but also in most cases restore or improve our gut microbiome.

I have started to think of myself as a hotel for bacteria, and emerging scientific research supports the idea that maintaining the well-being of the bacteria in our gut, mouth, and skin is key to achieving better health and happiness. These microorganisms influence not only our physical health but also our mood and cognitive function. Humans and these microbes have coevolved, and as we move farther away from our natural environments, our health tends to deteriorate. The disconnection from the land, seasonal cycles, growers, producers, and locally sourced food has already contributed to widespread health issues.

As hosts for bacteria, it is essential for us to provide the incoming microorganisms with good accommodations and to care for these guests by providing them with the right nutrients at the correct times. Every day there are new findings, and it is exhausting to keep up with them, but findings about our circadian rhythms and the gut microbiome shed light on the intricate relationship between our bodies and the ecosystems within them, as well as our position in time and space. Light sets our internal clock, and it is important to live in accordance with it to be healthy. Without delving into other environmental factors like indoor pollutants from artificial lights and off-gassing or even outdoor air and noise pollution, the nutritional value of local, seasonal produce stands out. Consuming plants when they are in their natural peak condition allows our bodies and gut bacteria to derive maximum benefits. For instance, eating a strawberry grown nearby and in good soil in the summer, when it is in season, is far more beneficial than consuming one in winter even if it is organic. The body knows, or has coevolved, and therefore the level and length of time of light in the environment informs it or the bacterial communities. This perspective aligns with a simple yet powerful principle: Eat and move as your healthy forebears did to maintain good health.

It is clear that a diet rich in diverse high-quality plants, animal products, and fermented foods, like yogurt, kefir, and fermented vegetables, can foster a healthy gut microbiome. As to how fermented foods affect our body, it is complex, and I'm not qualified to explain it fully. Simply put, it is mainly because fermented foods are a source of probiotics. These live microorganisms, whose name is derived from the Greek phrase meaning "for life," were first coined by German bacteriologist and food scientist Werner Kollath in the early 1950s, as cited in Vergin. He described *probiotika* as "active substances that are essential for healthy development of life."[3] Probiotics promote gut health by maintaining a balanced microbiome, which aids digestion, enhances nutrient absorption, and produces essential compounds like short-chain fatty acids (SCFAs). Additionally, probiotics bolster the immune system and protect against harmful pathogens by outcompeting them in the gut.

You may also have encountered the term *prebiotics* (before life), which are nondigestible food components, typically plant fibers or complex carbohydrates, that serve as food for beneficial gut bacteria. While they are not live microorganisms, prebiotics stimulate the growth and activity of probiotics, indirectly contributing to better gut health.

Postbiotics (after life) are the bioactive compounds produced by probiotics during fermentation and metabolic activity. These include SCFAs, enzymes, peptides, vitamins, organic acids, and other metabolites. Like prebiotics, postbiotics are also not live organisms; rather, they are by-products of probiotic acidity. Despite this, they can still provide many health benefits. Research is rapidly expanding on these topics, particularly the fields of gut health and longevity, and evidence continues to emerge about their potential role in supporting overall well-being.

Fermented foods in general and fermented dairy products, for those who can eat them, offer significant health advantages. As with all foods, but particularly fermented foods, it is always best to start with smaller portions, building up gradually. It only takes a short while for the gut bacteria to start processing and enjoying the newly introduced foods with live bacteria.

While dairy ferments are not suitable for everyone and certainly not a cure-all, their benefits are well documented and seem to grow daily. Yogurt in particular has been extensively studied and shown to be good for health. Similarly, more research is being done on the effects of other popular dairy ferments. Studies investigating the genes of milk, starter cultures, and the resulting ferments have provided fascinating insights into their effects on the body, particularly on the gut-brain axis.

There is also a growing interest in the role of protein in longevity, leading to recommendations for increased daily protein intake. The rising popularity of whey, kefir, and other fermented dairy products high in protein is evident in the expanding range of options in supermarkets. Each visit to the refrigerated section seems to reveal new fortified yogurt or protein-rich whey products. Unsurprisingly, some research has been misapplied, usually in an effort to boost sales or sway public opinion.

Here are a few health benefits of fermented dairy (as of publication):

- Fermented dairy products support a balanced gut microbiome, which is essential for effective digestion, immune function, and the prevention of gastrointestinal issues.
- Fermentation enhances the bioavailability of key nutrients like calcium, vitamin D, and B vitamins. These are critical for maintaining bone health, energy production, and overall wellness.
- Fermented foods like soured yogurt typically contain lower levels of sugar compared to cream-style yogurt and may be more suitable for individuals with metabolic disorders like diabetes.
- Certain strains of *Lactobacillus* and *Bifidobacterium* found in fermented dairy products can lower cholesterol levels by breaking down bile acids, thereby reducing cholesterol reabsorption and promoting its excretion.
- Chronic inflammation is a factor in many diseases, including obesity, diabetes, and heart disease. Fermented dairy products like kefir and yogurt have been shown to reduce inflammatory markers in healthy individuals and those with metabolic issues.
- Probiotics in fermented foods strengthen the immune system and inhibit harmful pathogens by competing for resources in the gut.
- Fermented dairy products rich in probiotics may positively influence brain function and mood through the gut-brain axis. They can produce neurotransmitters like serotonin and influence pathways that impact mood, behavior, and cognitive function. The latest research into psychobiotics explores how gut health influences emotional well-being.

While fermented dairy offers significant health benefits, it is not without risks. Low-quality dairy, poor animal husbandry, improper hygiene during production or storage, and contamination can lead to serious health hazards. Strict manufacturing protocols and proper regulation are critical to mitigate these risks and to ensure the safety and health benefits of fermented dairy products.

Here are a few additional concerns:

- Fermented dairy products can be high in biogenic amines like histamine, which may trigger migraines or inflammatory responses in susceptible individuals.
- High-fat cheese and other fermented dairy products with elevated saturated fat or sodium levels may increase low-density lipoprotein (LDL) cholesterol or raise blood pressure in certain individuals.
- Carcinogens like aflatoxin M1, growth hormones, and antibiotic residues in some dairy products can disrupt the endocrine system and compromise immunity.
- Finally, despite fermentation reducing lactose content, individuals with severe lactose intolerance or dairy allergies may experience bloating, abdominal pain, and other more serious health issues.

We know that these benefits only apply to people who can safely consume dairy. For those who can't, there are two main conditions. The first is lactose intolerance, a common digestive issue in which the body cannot break down lactose due to insufficient lactase enzymes. The second is dairy allergies; here the individual has an immune reaction to milk proteins like casein, leading to symptoms that range from mild (hives, nausea) to severe (anaphylaxis). These conditions are often misdiagnosed or self-diagnosed, leading to people avoiding dairy altogether.

I touched on this briefly in the first chapter. We see how history and biology intertwine, and it seems those humans who consumed processed dairy in their past tended to be more lactose intolerant than those who had consumed raw milk. This sometimes led to cultural snobbery, with some groups considering it prestigious to process dairy and expressing disdain for those drinking it fresh. Biologically, most people can digest milk as infants, but lactase production often drops after weaning. Those who retain lactase into adulthood, a condition called lactase persistence, can continue digesting milk. Theories vary about why lactase persistence occurs. The cultural-historical hypothesis suggests that our bodies adapted to dairy consumption over time, while the reverse-cause hypothesis proposes that dairy consumption became common among populations with lactase persistence. For example, Tibetans, who have a history of daily milk consumption, have higher lactose tolerance than some of their Chinese neighbors.

Relevant to this discussion is another method for diagnosing dairy-related digestive issues: the FODMAP (fermentable oligosaccharides, disaccharides, monosaccharides, and polyols) diet. It is primarily used to identify foods that trigger digestive symptoms in conditions like irritable bowel syndrome, including lactose intolerance. This is done by abstaining from certain food groups. For example, all carbohydrates (or sugars) may cause digestive discomfort for some people, and by systematically avoiding the foods, then reintroducing them, it is possible to find out what is causing their upset. Dairy products such as A1 cow's milk, condensed and evaporated milk, yogurt, ice cream, and custard have high levels of lactose, a disaccharide, considered a "high-FODMAP food." Once the cause of their allergy or intolerance is identified, those people who are cutting out lactose may be advised to have plant-based milk, like rice or lactose-free milk; Brie, Camembert, and Feta cheese; and some other hard cheese with no or low lactose as a substitute.

There is a whole industry centered on easing the symptoms of lactose intolerance with medications and supplements. In the commercial fermented food industry, researchers are searching for ways to reduce lactose-related digestive disorders by creating products such as synbiotics, which are mixtures of probiotics and prebiotics designed to improve gut functions or alter the composition of our microbiome for the purpose of improving lactose digestion.

Soon there may be fermented dairy products available with dosage instructions tailored to each consumer's taste and body composition. These "functional foods" are being studied and are an area of growth for industrial food production. Clinical research into fermented products, especially fermented dairy, will likely shape future nutrition and health programs. These studies might even pave the way for the next generation of therapies to prevent, manage, and treat diseases. Recent advances in AI and genetic sequencing are enabling the development of precise prebiotic, probiotic, and postbiotic formulations. These innovations aim to capture the benefits of fermented foods while mitigating any drawbacks, potentially leading to custom-made products tailored to individuals.

Furthermore, the use of genetic analysis in studying fermentation microorganisms has revolutionized the field. By isolating or combining specific traits and employing AI-generated models, researchers can optimize yields, qualities, and substrates. Such foods could potentially be produced quickly, efficiently, and affordably. While this approach differs significantly from traditional fermentation practices, it represents what some may consider progress. We will have to wait and see.

Supermarket dairy aisles are full of new fermented products with added protein or vitamins and minerals. These synthetic white liquids and pastes, often pasteurized and sold in colorful packaging, lack most of the health benefits that can be gained from consuming heirloom diversity, which is found in natural ferments.

Fermentation as a whole, including dairy fermentation, is a growing and significant sector in the global food industry. Beyond the familiar dairy products, as touched on above, milk components such as sugar, fat, and protein are increasingly used to fortify and enhance various foods. However, individuals with dairy intolerances or allergies must remain vigilant. For instance, lactose is often utilized as a sweetener or to improve browning and texture in mass-produced foods like pancakes, pastries, and other processed goods.

Fermentation, an ancient, naturally occurring process, has found numerous modern applications across industries such as pharmaceuticals, where fermentation is essential for producing antibiotics, vaccines, growth hormones, and insulin, among other medical innovations. In the chemical sector, fermentation contributes to the development of products like dyes and medicines. Additionally, fermentation has proven valuable in food-waste management and environmental remediation efforts. Whether used to develop new medicines or biofuels, fermentation necessitates advanced scientific and technological capabilities to create, measure, and analyze products. These prerequisites not only boost economic benefits but also expand our understanding of this foundational process.

Fermentation and fermented products also support agriculture. In Japan, for example, *bokashi*, fermented organic matter, improves soil health, while silage, a fermented cattle fodder, provides easily digestible nutrition for livestock and other animals.

As we have seen, ferments are made from a community of microorganisms, and some fermentation starters bring about human communities. One such is Herman the friendship cake sourdough starter, which I was given and used to make cakes over ten years ago. Those of us who had

Herman shared ideas and recipes via email. Thankfully, Herman didn't live too long in my care, as we were eating a lot of cakes.[4]

Fermented foods are cherished across the globe; with each home, village, and region producing unique batches, no single method or condition guarantees identical results. Home fermentation is as much an art as a science. Variables such as milk quality, ambient conditions, and the interaction of microorganisms all influence the outcome. Whether you are fermenting at home or purchasing artisanal products, experiment and learn from the process. Remember that fermentation is a collaborative effort between the maker and the microbes. Even when things go awry, it is often due to the complex "company" of microorganisms rather than individual error. Embrace the uniqueness of each ferment and enjoy the journey of discovery.

Despite advancements in sanitation and access to standardized ingredients, the process of fermenting dairy at home remains largely unchanged from ancestral practices. The quality of modern milk, influenced by environmental pollutants and farming practices, differs from the natural, unaltered milk of the past, which produced richer, arguably better and more diverse ferments.

The fermentation instructions given below can only serve as guidelines. To make fermented dairy at home, you do not require specialist equipment—only a good thermometer and, perhaps, your gut instinct (if you will pardon the pun). Trust in your intuition, which has guided humanity for millennia, even though modern life often encourages reliance on companies and experts who aren't always trustworthy. For instance, many of us obsess over sell-by and use-by dates, losing touch with the natural instincts that served our ancestors well in most cases.

Start with clean, nonreactive containers, anything from glass jars to enamel pots. Historically, people used a variety of vessels, from animal skins and clay pots to gourds and woven baskets, each offering unique properties. If using pottery, ensure it is free from harmful lead-based glazes.

Water quality is crucial. Tap water often contains chlorine, which inhibits fermentation, and even boiling may not remove other contaminants such as antibiotics, hormones, or "emerging pollutants." Some suggest using water that has been treated with activated charcoal or boiling water and letting it sit overnight to reduce chemicals, but this is not a perfect solution.

When adding salt to ferments, choose varieties without additives like anticaking agents or iodine. Natural salts with trace minerals can enhance flavor and nutrition. The amount used depends on personal preference and the specific ferment.

Fermentation microorganisms generally thrive away from direct sunlight, though ambient conditions such as temperature and humidity play a significant role. For instance, kefir left on a kitchen counter in the damp climate of the UK may require different care compared to one being made in an arid region.

Modern materials like stainless steel and thick glass can serve well in home fermenting. For fizzy ferments, thick glass bottles with secure lids are ideal. Plastics, however, are best avoided due to their tendency to leach chemicals into acidic ferments. Lately, I have been wondering if there are microplastics among other contaminants in the milk I use to ferment my kefir. Could this affect the grains? They are an ancient stock that I feel responsible for preserving. Strangely, I seem to care more about the impact on the grains than on myself.

If you are planning on adding fermented products to your diet, consider supporting local artisan producers who prioritize quality and traditional methods. Each ferment is unique, and finding ones that align with your taste and microbial preferences can be deeply rewarding for you, or should I say your bacteria?

4

Milk's Magical Metamorphosis

Raw milk begins to transform as soon as it leaves the animal. Historically, pastoral nomads addressed this challenge by taking only the milk they needed for immediate use, leaving the rest stored within the animal. Even today, in small villages in warmer regions, milkmen will often milk animals directly at the customer's doorstep, ensuring the milk is fresh.

Currently, there is a resurgence of interest in raw milk, available from local farms, as well as unpasteurized and fermented dairy products found more widely. While modern practices emphasize the sterilized handling of milk, historical practices involved milking with relatively clean hands into relatively clean vessels washed with water. Before the availability of hot water on tap and dishwashing products, these methods sufficed to remove most contaminants.

Milk, however, was still exposed to accidental contamination from microscopic organisms and physical debris, including insects or even animal excretions, which might dissolve unnoticed into the warm liquid. During a visit to Lighvān, a village famed for its ewe's milk cheese, I observed ladies using narrow-necked clay jars, known as *chuza* in Āzari, when milking. It seemed this narrow opening would make the process unnecessarily awkward, but when I asked, I was told it served to reduce contamination!

In previous chapters, we saw how the vessel used to contain milk could also affect the milk in different ways. For example, storing milk in the stomach of an unweaned ruminant, where the enzyme chymosin is present, causes the milk to turn into cheese. Alternatively, when certain bacteria predominate, yogurt forms, while a combination of bacteria and yeast produces kefir. In the case of *chortān* and *gooroot*, curds were dried and could even be stored for months or years when handled properly. Each fermentation process produced distinct textures, flavors, and nutritional profiles. Fermentation made milk, a perishable and difficult-to-transport food, more durable and portable.

All the transformations mentioned above begin with clabber, which occurs when whole raw milk is left undisturbed in a container. First, the fat rises to form a layer of cream, while the milk remains beneath. Then the natural bacteria present in the milk, cream, and the surrounding environment initiate changes. I would like to start with clabber in its own right. It is the simplest and most natural of the dairy ferments, and it just happens. We have various phrases for this as I have described in the previous chapter: We say the milk has clabbered, gone off, curdled, or soured. Nowadays, clabbered milk is often discarded as unusable, though our ancestors used clabbered milk and cream. If the milk had clabbered so much that there were curds and whey, these too would have been eaten and drunk.

To get clabber, we need milk. What is interesting is that if I were writing this one hundred years ago, I would just say *milk*—in fact, throughout the whole book—but now I have to say "natural," "raw," or "organic" milk. The same applies to my writing on food as well. I find this quite sad and, in a way, divisive, because this type of "natural" food is more expensive and less accessible to most people.

The use of whey from all ferments in this book is covered in the final chapter, which focuses specifically on whey and other liquid ferments.

To begin, we pour four liters (1 gallon) of whole natural milk, straight from a healthy animal, into a glass container and leave it at room temperature. For me, in the UK, this is (on average) around 17°C (63°F).

- In the first few hours, you will notice the cream rising to the top, and there will be milk beneath it.
- At this stage, you can skim off the thick cream, which is known as heavy cream in the United States.
- After a little while, you can skim it again, and you will get a thinner cream for pouring, similar to single cream in the UK and half and half in the United States.

The cream can be eaten, beaten to make sweet butter, or used to fry food—for example, eggs.

What is now left in the container is milk, which (if you can digest milk) you can drink au naturel or flavored, and it can be made into milkshakes or chāi lattes. Alternatively, you can use the milk as an ingredient in a variety of recipes, such as sauces, poaching liquids, and baked goods, among many other things.

Another less popular way to consume milk is to cook it slowly, either on the stovetop or in the oven in a wide, shallow pan. The reduced milk develops a caramelized flavor and color and becomes thicker in texture. It smells wonderful too.

So far, the milk and cream have been fresh and have not yet soured (that is, the bacteria have not yet converted lactose into lactic acid), but if they are left a bit longer, this will happen. In a warm environment, this happens more quickly than in cooler environments. As we saw in chapter 3, it is known as spontaneous or wild fermentation and happens because of the presence of microbes in milk and in whatever it comes into contact with. The resulting product is known as clabber.

Once the milk has clabbered, you can consume it like yogurt, either plain or flavored. Or you can strain the clabber, giving very soft curds and whey. These curds can be eaten plain or with added flavorings.

Similarly, the cream skimmed earlier, if left out, will also start to sour and become similar to sour cream. If you beat this soured cream, you get a cultured butter.

Throughout the process, you have to use your senses, and if at any point you feel the milk, cream, or clabber looks or smells off, then do not eat it; throw it away. When I first asked, "How will I know the difference between spontaneously fermented milk and spoiled milk?" I was told, "You will know." And yes, I did. Spoiled milk is quite off-putting, and I do not like to use the term *disgusting*, but it is. On the other hand, clabbered or wild fermented milk has a sweet, appealing, sour aroma. You are going to have to trust me and your instincts on this.

5

Which Came First, the Yogurt or the Pot?

Perhaps the most globally recognized and widely enjoyed of Central Asia's fermented dairy products is yogurt, or *yo-ud*, as it is pronounced in Āzari. Like many other fermented dairy foods, yogurt was likely discovered by accident through the natural processes of time and environment. I would like to start this chapter by sharing my favorite dairy-related myth, which relates to yogurt and may be about the possible origin of yogurt. There is a Greek myth that connects milk to the stars. The word *galaxy* comes from the Greek word *gala*, meaning "milk." According to this myth, the milk of the goddess Hera (protector of women and childbirth) spilled into the heavens, creating the stars of the Milky Way.

It also happens that in Greek, snowdrops are known as *Galanthus*, or "milk flowers," that are native to many parts of Eurasia and particularly the Caucasus. Analysis of the genome of yogurt bacteria indicates that it may have originated on plant surfaces, which have yet to be identified. I found one paper suggesting that snowdrops may have been the original plants for creating yogurt. I was so charmed by the story of snowdrop genetics in yogurt that I decided to jump to a conclusion of my own, because it is meaningful to me, a yogurt-loving, galanthophile Āzari.

> **Scenario 2**
> I imagine a young girl in the Caucasus during snowdrop season, which coincides with lambing in the region. Perhaps she had picked snowdrops; maybe one fell into the container of milk or the animal's udders brushed against the flowers, introducing their genes into the milk. The milk, exposed to additional microorganisms from the air and the container, transformed in consistency and aroma. The next day, the girl tasted it, found it safe and pleasant to eat, and shared it. Over time, the same container might have been reused, repeatedly inoculated with the same bacteria, which became established and stable. Perhaps this container then became the preferred vessel for this new food, and yogurt was born.

For millennia, yogurt-style fermented dairy has been made and enjoyed worldwide; wherever it has been introduced, people have embraced it, savoring it plain or flavored and incorporating it into their diet. As I pointed out in chapter 2, it is documented that Iranians consumed acidulated milk with honey. In the first century, the Greek physician Galen mentioned a popular dish known as *oxygala*, a dish combining possibly yogurt and honey. This may have come to Greece when Alexander III of Macedon (known not as "The Great" but rather as "The Destructive" from an Irani perspective) "visited" the region and picked up this food.

Of all the ferments, I would say yogurt is the most researched and written-about dairy product. There are hundreds of books and research papers on yogurt in different languages. Some of the research focuses on its etymology, microbial genetics, archaeological findings, nutritional value, health benefits, and industrial and culinary applications. Scientists, anthropologists, medical professionals, Nobel laureates, and chefs have all explored its many dimensions. Similarly, there are many different recipes for making yogurt or using it as an ingredient.

In 1905, Bulgarian microbiologist Stamen Grigorov identified the bacteria necessary for yogurt fermentation in his homeland, naming them *Lactobacillus bulgaricus* and *Streptococcus thermophilus*. Since then, these discoveries have set the standard for most yogurt production and influenced regulatory guidelines. Besides these two, numerous other bacterial strains and local microorganisms contribute to the diverse types of yogurt produced worldwide. These varieties differ in texture, flavor, and mouthfeel. While some yogurt are firmly set, others are softer. Some are tart, while others are creamy.

However, today, industrially produced yogurt can only carry the name "yogurt" if it meets certain standards. It is formally described as the coagulated milk product created by the lactic acid fermentation of milk using specific bacteria: *Lactobacillus delbrueckii* subsp. *bulgaricus* and *Streptococcus thermophilus*. The microorganisms in the final product must be viable (alive) and plentiful.

The first mass-produced yogurt was sold by Danone in Spain, founded by Isaac Carasso in 1919. Another well-known yogurt brand in the United States is Chobani, started by Hamdi Ulukaya, a Turk from Kurdistan. The name Chobani, meaning "shepherd" in Turkic, reflects its cultural roots. Today, there are rows of plain and flavored yogurt in many supermarkets around the world.

These types of yogurt are usually made by fermenting mass-produced cow's milk with defined strains of bacteria in a sterile environment. They may also be thickened and flavored and have preservatives added to create products with a longer shelf life and characteristics that are best suited to a particular nation's taste buds. Much like mass-produced sourdough bread, industrially made yogurt doesn't always get the quality of ingredients and particularly the time it needs to naturally develop its unique qualities.

Traditionally, the only addition to yogurt was salt, which enhanced the flavor and extended shelf life depending on the amount used. This may have been done for millennia, as evidence suggests that salt processing dates back thousands of years, and there was an increase in salt use in agricultural settlements, serving both dietary and preservative purposes. Nomadic lifestyles provided salt naturally through diets rich in dairy, meat, and blood. Central Asia and the Caucasus, with their salt lakes, mineral-rich rivers, and mountainous rock salt deposits, offered abundant sources. In high mountain regions, where it is cooler, less salt was needed, while in the hotter, lower-altitude plains, more salt was used to extend shelf life. This also applies to preserving dairy. Yogurt made for long-term storage is generally saltier, such as winter yogurt, or *gish yogurdi* in Turkic.

Today, there are many products called yogurt that resemble it in texture but not in microbial composition, such as those curdled with chili calyx, tamarind, kefir, or leftover whey from other dairy. Products like skyr, which is an Icelandic low-fat cheese, sometimes called Icelandic yogurt, are often classified differently and cannot officially be called yogurt.

Let us now look at its name. The word *yogurt* or similar sounding words are used in many countries, particularly those influenced by Osmani culture in Eurasia and the Americas. For example, in my adopted country of Wales, it is called *iogwrt*. Other influences have affected its name elsewhere: It is known as *matsoni* in Georgian and *mast* in Fārsi.

I do not think we will ever fully agree on a single pronunciation or spelling for what is called *yogurt* in the United States, but it is fun to see how passionate people are about it. I spell it "yoghurt," as I grew up in the UK. Regardless of how you spell it, the word is undeniably of Turkic origin. There are several theories regarding the meaning of this word. The most common meanings are "kneading" and "coagulating," both derived from the Turkic verb *yoghurmak*. Other interpretations, like "to make it sleep," are also plausible. To me, it makes sense that the milk is "sleeping," or *yātmokh* in Āzari. When yogurt is incubating, any experienced yogurt maker will make sure it remains cozy and undisturbed, rather like taking a nap.

Should we let sleeping yogurt lie? Why? I have another word for *yogurt* for you. In Āzari and Armenia yogurt is also known as *gātikh*. In Turkic languages, the verb *gāt* means "to mix," and *gātikh* is "a mixed thing." I would like to suggest that this name was given because you mix in some yogurt from the previous batch to make yogurt, what is also known as backslopping.

The intricate etymological journey of *yogurt* and *gātikh* reveals a deeper story of cultural and linguistic exchange. These words, born in the heart of Eurasia, migrated, adapting and branching out like the nomadic communities that carried them. They stand as a testament to the enduring power of food and language in shaping our shared culinary heritage and the rich history and cultural diversity surrounding this ubiquitous dairy product.

Yogurt holds such cultural significance in many Central Asian cultures; in Iranian society, it even appears in poetry and common expressions. For instance, in his *Golestan* (*The Rose Garden*), Saadi warned that yogurt from a stranger should be avoided, as it might be more than two-thirds water. The term *mast mali kardan* is used to describe someone who is making a situation unnecessarily complicated, much like spreading yogurt everywhere.

In her book *The Legendary Cuisine of Persia*, Iranian cookbook author Margaret Shaida recounts how a popular cold yogurt dish, *borāni*, got its name. The story goes that it was one of the favorite meals of Burāndokht (or Pourāndokht), the first queen of the Sasanian Empire in the seventh century, and that this dish was later named after her.

Yogurt is a staple and is present at most meals in Central Asia, either as the meal itself or as a side dish—for example, in Iran, with *khurush* or *āsh*. We have it as a snack with bread, and it is also used as an ingredient in marinades, bread making, cakes, and cold dishes like *ābdoogh* (a cold soup). Yogurt with jam and honey, as mentioned above, is a popular dessert and likely the origin of many modern yogurt "corners," packaged yogurt with a fold-over corner containing sweetened "fruit," sold as a healthy snack.

Yogurt has been appreciated for its health benefits for centuries. In regional folk medicine, it is often seen as a mild food for those who are unwell and has also been used topically to soothe bee stings or sunburn. Changiz Khan, the founder of the Mongol Empire, is said to have fed his army a variety of dairy products to give them strength; perhaps this included yogurt. In the sixteenth century, King Francis I of France introduced yogurt to Europe after receiving it from his Ottoman allies as a remedy for diarrhea.

In the early twentieth century, Nobel laureate and zoologist Ilya Metchnikoff's research on probiotics helped popularize yogurt as a health food. There is extensive research on the health benefits of natural yogurt, including being a great source of probiotics. Fermented dairy products, including yogurt, are said to reduce the risk of certain cancers, improve weight management, and support cardiovascular, bone, and gut health. However, the fat, sugar (lactose), and salt in yogurt may be harmful to those with certain medical conditions. As we have seen, most naturally fermented foods develop their flavors and benefits over time.

Nowadays, aside from salt, synthetic preservatives, colorants, acidity regulators, vitamins, minerals, protein, and sweeteners are commonly added to create new "functional foods" or "health" products and to align with the latest food trends. Some of these marketing labels add to the confusion with terms like *live, cultured, probiotic,* and more recently, *protein*. In general, well-made yogurt remains one of the healthiest foods when enjoyed in moderation.

Natural yogurt contains macro- and micronutrients and is a great source of probiotics. As people become more aware of food intolerances, allergies, and the gut microbiome, demand for fermented foods like yogurt, with their associated health benefits, is on the rise.

To prepare natural plain yogurt, a small portion of a previous batch is added to warmed milk to initiate fermentation; as mentioned above, this is commonly known as backslopping. The mixture is then left undisturbed

Yogurt

in a warm location until it sets. The resulting yogurt is creamy, neither runny like single cream nor solid like butter. Its flavor is mild and creamy, becoming tangier with time. Yogurt is made with ewe, nanny, buffalo, dri, and camel cow's milk throughout Central Asia. Now almost all industrial yogurt is made with cow's milk. It is made with full-fat raw, pasteurized, and UHT milk. Interestingly, some individuals achieve better results when making yogurt with pasteurized or industrially processed milk, as the bacteria in the starter culture do not have to compete with naturally occurring microorganisms.

Making plain yogurt at home is simple, requiring only milk, yogurt, and salt. A thermometer is helpful for beginners, as maintaining the correct temperature is crucial. Other equipment includes a pot, a mixing bowl, lidded containers for storage, one small jar, and an incubator of sorts to maintain the right environment for fermentation. Incubators can range from a cardboard box lined with a blanket, many layers of warmed towels, low-temperature ovens, and heat pads (like ones used to germinate plants or to warm a bed) to airing cupboards in traditional UK homes.

How to Make Simple Plain Yogurt

Note: Be gentle throughout the process.

Ingredients

1 L (34 fl. oz.) of raw milk (you can also make it with pasteurized, UHT milk as well)*
2–3 tbsp. of plain live yogurt (either shop-bought or from a previous batch), known as the starter
1/2–1 1/2 tsp. of salt (adjust to your taste)
A small jar

If using raw milk, heat to 90°C (194°F) for twenty minutes to pasteurize it. Avoid overheating or boiling it rapidly to prevent curdling. *Tip:* If you like thick yogurt, you can do the heating process slowly so some of the water in the milk has a chance to evaporate out. This step neutralizes some of the strong bacteria in raw milk that might overpower the yogurt culture, and it also denatures the protein so it sets.

Cool the milk to around 43°C (110°F) before adding the starter.

Mix the starter into one cup of warmed milk so that it is uniform.

Gently pour this mixture into the rest of the heated milk and stir it in slowly.

Then incubate the mix by keeping it warm and cozy.

There are many ways to incubate the milk, but here is what I do for a large batch or several small jars.

For a large batch of yogurt made in a bowl, place a large colander or sieve upside down over the bowl to create a dome over it, then cover it with several layers of thick cloth or a blanket. Leave it undisturbed in a draft-free place at around 35°C–40°C (95°F–104°F) for eight to ten hours for the yogurt to set. When I last bought an oven, I chose one that started at 35°C (95°F) and included a nonfan function for minimum disturbance.

Alternatively, if I'm making small jars of yogurt, then I warm the jars in the oven on its lowest setting or pour boiling water into the jars,

---

* If using pasteurized or UHT milk, just bring it up to 43°C (110°F).

> then tip it out and fill them with the mixture. Either I use the oven or I line a cardboard box with a blanket, put the jars in, and close the box. In the winter, I wrap another blanket over the box and, as above, leave it undisturbed for the duration.

After you have made the yogurt for the first time, you can start adjusting the salt and length of incubation to get the flavor and texture you like. The longer it incubates, the more sour it becomes. If you want your yogurt to be more set or tangier, let it incubate a little while longer, maybe a day or two.

Once it reaches the desired acidity and consistency, transfer it to the fridge. Stored in the fridge, it should last for a couple of weeks. For thicker yogurt, *suzma* (strained) or *torbā gatukh* (sack yogurt) in Turkic languages, known as "Greek yogurt" in the West, pour it into a cloth-lined sieve and allow it to strain. Do not throw the whey away. With experience, you will learn to judge the set, perhaps by observing its jiggle, as my friend Fariba's mother does.

You too will develop a routine and adapt the process to suit your preferences. Yogurt making, like all fermentation processes, is both an art and a science, a blend of tradition and personal experimentation. One day you will have thin yogurt; another day, it will be thick. That is just how it is!

The small jar I mentioned earlier is essential, as it is where you store the starter for your next batch. I always save some after making yogurt. You can buy a good-quality plain yogurt and use that as your starter.

You can also buy or make your own products similar to yogurt using plant-based milk with a starter and the same method described above.

Depending on the milk and starter you use—for example, if it is nut, seed, or grain based—you will need to experiment to get the right set and flavor. Some plant milks take longer to set and are more difficult to work with, while others set easily and are more straightforward to use. The outcomes vary depending on the brand and type of plant-based milk and starter. I do not have any names to suggest for this type of product. I read that it is sometimes referred to as soygurt, especially when made with soy milk. I haven't used this term; I've said dairy-free "yogurt" in quotation marks. Besides, you can use any plant-based milk to make it.

You may have come across, or tasted, what is known in the West by the Arabic word *labna*.[1] It is made by straining yogurt overnight until it becomes thick enough to be rolled into balls. These are often also referred to as cream cheese and enjoyed in much the same way: plain, sweet, or savory. Savory *labna* is often preserved in oil. In the next chapter, we see how these balls can be dried into *chortān*.

Dairy-Free "Yogurt"

Note: Be gentle throughout the process.

Ingredients

1 L (2 pt.) of plant-based milk
2–3 tbsp. of plain plant-based yogurt (either shop-bought or from a previous batch), known as the starter
½–1½ tsp. of salt (adjust to your taste)
A small jar

Heat the milk to around 43°C (110°F) before adding the starter.
 Mix the starter into one cup of warmed milk so that it is uniform.
 Gently pour this mixture into the rest of the heated milk and stir it in slowly.
 Then incubate the mix by keeping it warm and cozy. Follow the instructions above for incubation, storage, and making *suzma*. If you make it, call it something else, as it isn't dairy.

I have a deep personal fondness for plain yogurt, one of the first foods I remember eating as a child. I prefer it sour, so I leave shop-bought, unopened pots at room temperature for a few days before opening them to use; then I store them in the fridge. If I go without yogurt for a few days, I begin to crave it, or is it my bacteria who are craving it?

When I was a child, my father would buy yogurt from his friend Mr. Bahman, who was a third-generation skilled yogurt maker. For our personal use, twice a week, he'd bring home a one-kilogram (2.2 lb.) *seyin* (a shallow, wide, earthenware dish). If we had a dinner party, he'd buy a *matrat*, which is a large *seyin* that can hold four kilograms (8.8 lb.). The pots were covered in a crinkly skin (lactoderm) that only my father enjoyed. Most evenings before supper, we'd snack on yogurt with bread, using flatbread triangles folded into tiny cones to scoop it up. My father always had three cones for himself and gave me one, sometimes testing whether I noticed that he had cheated me out of my share. I always did. If my mother ever asked him what he wanted for dinner, his answer was invariably *youd chorayh*, or yogurt and bread.

Buffalo cow's milk yogurt with lactoderm pushed back

A few years ago, I happened to pass Mr. Bahman's shop and stepped inside to speak with the man behind the counter. It was Mr. Bahman's son. When I mentioned that my father used to buy yogurt from his father, we both became emotional. It was a moving moment brought about by our shared connection to yogurt. He even brought out an old glazed *seyin* for me to see. As I've said before, it isn't merely yogurt, a

Mr. Bahman in his yogurt shop

simple food; it is bound up in memories, belonging, and for me, a taste of home.

When my family first arrived in the UK in the 1970s, finding plain yogurt was a challenge. The only option we found was a brand called Ski, which was pink and strawberry flavored. For us, yogurt was a daily staple, while in the UK, it was considered either a dessert or a health food. Eventually, when we found some plain yogurt, my mother started making our own. It was precious, and she took care of the starter, much as our ancestors likely did.

I began this chapter with a story about the possible origins of yogurt, and I'd like to end it with *amasi*, a South African fermented dairy product similar to yogurt. I remember reading about it in *Long Walk to Freedom* by Nelson Mandela. *Amasi* is a fermented cow's milk product made in a gourd or calabash, which he made and enjoyed. Making it nearly gave away his hiding place. He had been leaving it to ferment on a windowsill, and it was noticed by a couple of Zulu workers, whom he overheard commenting on how unusual it was to see it in a white area. Needless to say, he stopped making it. By now it might be obvious to you that I, too, would have gone to great lengths to make sure there was yogurt in my life, as I feel quite insecure if we don't have yogurt, cheese, bread, and eggs in the house.

If you decide to make your own yogurt and you are heading to a party, you could take a jar of it, either to enjoy there or for it to be used as a starter. Add a label with instructions. Sharing starters is like passing on recipes or plants: Every time we see these gifts, we are reminded of the giver.

6

Gooroot: Dried Dairy Balls

Oh, Lord, this is not just a stone
I could smell the scent of milk from them

In this chapter we turn our attention to dried dairy—namely, *gooroot* and *chortān*, which are concentrated, protein-packed, calcium-rich "bombs," as I like to call them. For me, the most ingenious use of milk is *gooroot*, made with leftover whey. It stands as the ultimate expression of thrift, a tradition that ensures not a single drop is wasted. This is a food born of necessity that requires time as an important ingredient. For example, it takes time to make *gooroot*, to dry it, and for the flavor to develop, and then it takes time to rehydrate it before use. Being preserved, it also lasts for a long time. I am reminded of Harvey Mackay's quote "Time is free, but priceless": Unhurried and made in its own time, *gooroot* brings an "animally funk" to the *sufra* (tablecloth).

Drying food is among the earliest methods of preservation, enabling foods to last longer while becoming smaller, lighter, and more portable. In arid parts of Eurasia, people have utilized the sun and dry conditions to dry foods for millennia. These preservation methods allowed humanity to thrive by offering nutrition during travel and lean seasons, trade opportunities, and a basis for future dairy products. Unwieldy curds and whey, without refrigeration or storage solutions, would not endure the *koch*, or seasonal nomadic migrations. Large gluts of milk in one season were transformed into nutritious, convenient food that could be consumed all year round, at home and on the go.

Gooroot and/or *chortān* are important to the history of the region. They were part of the food that sustained the Mongols on their conquests of the region and are still popular throughout Central Asia. They have traveled across the earth and into space with cosmonauts from Baikonur Cosmodrome. They may have been threaded on a necklace to take into the afterlife. To some, they are a symbol of resilience, inspiring art, and are also used in rituals. Despite their rich cultural and historical significance, they are virtually unknown in the West—except for the brief TikTok trend "dry yogurt." It does not appear on many cookery websites or shows and is not on the celebrity chefs' ingredient lists yet. I would like to be for *gooroot* what Delia Smith (cookbook author and TV cookery teacher) was for cranberries, as we say in the UK. I hope the "Simi Effect" will bring *gooroot* into kitchens everywhere.[1]

Scenario 3

So far in my imagined story of dairy fermentation throughout the book, a young woman has been collecting milk from her animals in the usual vessel, inoculated with snowdrop bacteria to make yogurt, knowingly or unknowingly. It is plausible that her containers might have been leaky. Upon returning to the stored milk, she might have found curdled solids at the bottom of the dish. Perhaps she enjoyed them and, as they were safe to eat, made them again. Maybe she even started using leaves with small holes, woven baskets, or other cracked pots to separate the curds and whey. Maybe she also placed additional containers underneath these to collect the whey.

Now, if these curds were accidentally left out for a long while in the summertime, she might have returned to find that they had dried out. She may have initially been worried, thinking they were ruined, but after eating a piece, she may have found them palatable and noticed they lasted longer. Collecting them, she would have noticed that they were lighter and easier to transport, and so we get a new dairy product: *chortān*.

At first, these may have been eaten dry, until one day, a piece became wet, due to either rain or falling into water. It was rehydrated and became a drink. Perhaps another time a piece fell into milk and caused it to clabber, so it was then used as a starter culture. In the same way, some herbs may have fallen in, giving it a pleasing flavor, and thus soup was born. Later, eating this with bread or grains turned it into a more substantial meal.

The scenarios I have made up and my attribution of these discoveries and innovations to women align with the likelihood that these early societies had basic divisions of labor to ensure survival. As I explained in the introduction to this book, women were, and often still are, commonly associated with feeding and nurturing roles, which would have included tasks like collecting, storing, and preparing food for their groups. These roles underscore the critical contributions of women to the development of early food practices and the survival of their communities worldwide.

Before we delve into the topic, I have to point out that in most of the literature I found on dried preserved dairy, it was referred to exclusively as *gooroot* (with a variety of transliterations and romanizations, which we will see later). *Chortān* was not mentioned at all. I'm not entirely sure why, but as an Āzari, I make this distinction because we differentiate between these different but similar products, which I describe later. Having said this, even in Āzarbāijān *gooroot* is better known and used than *chortān*. *Chortān* is usually homemade, and I've never seen it for sale.

Let us now turn to how *gooroot* and *chortān* are made. Both can be made with a variety of milk, yogurt, or whey at various stages of dairy processing. I will start with milk that has clabbered. In my area of Āzarbāijān, if you strain clabbered whole milk, press the curds to get the moisture out (*lor*), and then salt, roll, and dry them or do the same thing with strained yogurt or *suzma*, these two types of dried dairy are called *chortān*, not *gooroot*. *Chortān* retains most of its nutrition, as it has not undergone any heat treatment.

So what is *gooroot*? For an Āzari, *gooroot* is dried dairy extracted from cultured butter whey. To make it, first yogurt is made with whole milk. The yogurt is churned to make cultured butter, and the remaining whey/buttermilk is turned into *gooroot*. This leftover whey is boiled, and small white solids start to appear, which are denatured whey proteins. These are collected by pouring the whey through a fine cloth and leaving it to strain. The runoff whey from this process is collected for the next iteration of dairy preservation detailed in the following chapter. These curds resemble cottage cheese but are quite sour and are also known as *lor*, or sometimes *shor* (salty). To turn *lor* into *gooroot*, salt is added and the curds are kneaded into a uniform paste. The mixture is shaped into balls or patties and left to air dry, typically outdoors, out of the sun, covered with light gauze to allow airflow while keeping dust and insects away.

The size and shape of *gooroot* are also important. Depending on the maker, *gooroot* might be piped, pressed in pretty molds, rolled, extruded out of a mincer, or grabbed in the palm of the hand with finger marks evident. It is important that the balls are not too big (the largest I have seen were the size of a medium chicken egg), as they need to dry to the core. Besides, the larger they are, the longer they take to dissolve when rehydrating before use.

Gooroot and *chortān* are made when the weather is dry, which is usually in the late spring and early summertime, depending on the elevation and yearly climate. They are air-dried on woven baskets so that air circulates around them and they dry out evenly. Time is also a factor; if they remain moist for too long, bacteria will start to grow inside, and they will become musty.

Today in Tabriz, it is hard to buy this type of traditional dry *gooroot*. Instead, there are rows of industrially produced and packaged *gooroot* in paste form, already rehydrated, pasteurized, and ready for use. These products are labeled as *kashk*, the Fārsi word for *gooroot*; as you may recall, Āzari is not a written language. However, in Kazakhstan it is labeled *qurt*, which is the name used there.

Of the many *gooroot* makers and dairy specialists I spoke to while researching this book, Mr. Sadaghian was extremely knowledgeable and enthusiastic. He said he only sold Kurdish *gooroot* especially from the city of Mahabad, which is made with ewe's milk, because Tabrizis are famously very selective about the quality of their food. This type of *gooroot* is slightly yellow in color when dry but becomes white when dissolved in water and has a milky and morish flavor, making it highly prized in Āzarbāijān. He considered every other *gooroot* inferior and said lesser-quality *gooroot* was sold in other areas of Iran for export and was also used by large industrial *gooroot* manufacturers.

Mr. Sadaghian felt strongly that if *gooroot* was not made with milk from Afshari ewe's milk yogurt and dried in the open air, it was not proper *gooroot*.[2] He believed the taste and goodness of sunshine were captured in traditional *gooroot*, going so far as to claim it had higher levels of vitamin D. He would not have approved of my first attempt at making *gooroot*, which was actually *chortān*, when I dried yogurt balls in the oven.

Another interesting detail I learned from Mr. Sadeghian was that it was important to salt the strained curds using finely ground rock salt. He explained that other types of salt cause *gooroot* to crack or even explode

Mr. Sadaghian

during drying. In this region there is a variety of salt, from mineral-rich mountains, caves, springs, and lakes. These salts contain many minerals, including sulfites, which might be why he emphasized using finely ground rock salt that dissolves quickly. Furthermore, the quantity of added salt varies by climate; generally, more salt is added at higher altitudes to preserve *gooroot* in colder and wetter environments.

Salt not only preserves but also enhances the flavor of *gooroot*. Besides salt, I had not seen flavored or colored *gooroot* in Iran, except for special batches seasoned with herbs like thyme or summer savory (*Satureja hortensis*). However, in Uzbekistan, Kazakhstan, and throughout Central Asia, beautiful mounds of colorful dried dairy are displayed in the bazaars. They are also used in weddings and parties to decorate tables. Some of these *chortān* or *gooroot* are handmade and artisanal, but most are produced on a commercial scale. My friends Celia and Slavenka visited Kazakhstan while I was writing this book and came back with *qurt* as souvenirs. These included vacuum-packed, colorful balls flavored with spices, herbs, and fruit, small enough to fit in the mouth, like gobstoppers or jawbreakers.

Modern preservation techniques mean you can now get low-salt and softer versions. Our most recent find is little flat disks of *gooroot*, which are like chips. My friend Mina discovered these, and since then my family has been buying them regularly.

Similarly, traditional air-drying has been replaced with advanced dehydration methods and additives that are used to preserve and enhance flavor. These vacuum-packed, chalky, usually pasteurized balls and pastes often lack the robust flavor or nutrients of traditional, high-quality *gooroot*, made slowly and naturally.

In Iran, industrial mass-produced *kashk* is made in factories using *gooroot* balls like those discussed above. The balls are dissolved in water using machines, which make an almighty racket. Then the resulting thick paste is pasteurized, and binders and preservatives are added before the finished product is packed into jars, plastic pots, or pouches with a very long shelf life. This *gooroot* is labeled in English as "Whey" or "Persian Sauce." In Fārsi, it always says "Kashk." The ingredient list on these packages varies, but most list "dairy whey powder" or "whey and salt." Some add flour to bulk it out, though this is not always listed. Celiacs beware!

All Iranian groceries, including little corner shops in Tabriz, stock various brands of industrial *gooroot*. Some of these are also available in the UK and the United States. In the last fifty years, especially following the Iranian Revolution and the ensuing emigration, this has also become part of *gooroot*'s journey. Artisanal and industrial versions can be bought in specialist "Middle Eastern" shops and online.

We always use traditional dried *gooroot* with care. For example, we are advised to always check the balls by cracking them open before rehydrating to look for *mur* or "moths," the harmless exoskeletons of insects that get into the *gooroot* when fresh and die there, which, according to Mr. Sadeghian, is not such a bad thing. It could be seen as proof that they were organic!

I am reminded of two other insect stories relating to dairy, which I will share here; one is in an article about the cheese of the pharaohs. *Mish* is an old Egyptian cheese eaten over five thousand years ago. It is made by fermenting drained cheese in sour milk or whey for months or years. It is salty and sharp and has an ammoniac smell. The insects in this cheese inspired the proverb "The worms of the mish arise from it," an Egyptian way of throwing up your hands and saying, "Hey, what ya gonna do?"[3]

Gooroot: Dried Dairy Balls

Gooroot dried in different shapes

Then there are cheese mites, which are actually arachnids with a hairy bottom like Tyrophagus casei and Acarus. They thrive on mold found in and on aged cheese with natural rinds. Some cheese is mite ripened, such as the German Milbenkäse. There are many examples of critters in, on, and making our food—pollinators are a great example—so we should view these mutualistic stowaways in the same way.

In order to use *gooroot* balls, it is important to rehydrate them in a very particular way. In Turkic, we call this *gooroot azmākh*, or "*gooroot* crushing," but it is not actually about crushing dry *gooroot* into powder. Rather, it is about dissolving it gradually in water by rolling it around a dish until a thick,

A variety of packaged *gooroot*

cream-like consistency is reached. You must not powder the *gooroot* balls before mixing them into water; if this is done, the mixture will separate, and the *gooroot* will sink to the bottom like sediment, with the water on top.

I remember the sound of *gooroot* more than the dishes my mother would make with it. She would sit on the floor with a *seyin* (a shallow, wide, earthenware dish that is rough on the inside, similar to a Japanese *suribachi*) in front of her, rolling the *gooroot* in the bowl with her hand one way, then the other. The sound was a rhythmic, soothing noise, almost like waves washing over large pebbles on a beach. It was a labor of love and hard work.

We didn't have one, but a *gooroot ali* (*gooroot* hand) can be used instead of your own hand to roll the balls in water. In Fārsi, it is called a *kashkmāl* or *kashksāb* (*kashk* spreader or rubber). I was very excited to find a *gooroot ali* in the bazaar in Tabriz, as I had never seen one before. I did not buy it, but the stallholder gave me permission to photograph it—I think because he was bemused by my enthusiasm and happiness at seeing it. When I got home, my mother thought it was a gimmick. With it being square, she said, rightly, that you would not be able to roll it around a round bowl.

One hardly ever hears about people dissolving *gooroot* at home nowadays; instead, people—including me—buy it in paste form. In Iran, you can buy traditional *gooroot* already made into a paste from specialist dairy shops. It can be refrigerated for a month or frozen to keep for longer. Mr. Sadeghian gave me some he had made, and it was the nicest tasting *gooroot* we've had in years.

Finally, if you have dissolved *gooroot* balls by hand, you may be wondering what happens when you reach the consistency and quantity of *gooroot* you need but have leftover damp balls. These can be rinsed and left to air-dry again for future use. Nothing is wasted.

As for the flavor of *gooroot*, ewe's and nanny's milk, being high in caproic acid (also known as hexanoic acid),[4] gives dairy from these animals a

Seyin

Gooroot ali

certain character, or what I have called "animally funk" from day one of my cookery classes. Fermented dairy products are mild when fresh, but as they mature, the flavors deepen, and they become funkier. Some have described the smell or taste of *gooroot* as the essence of the animal whose milk it came from. I have not eaten those types of *gooroot*, so I cannot comment, but I understand how people presented with strong-smelling, unfamiliar food may describe it in unfavorable terms, just like with natto or fermented shark.

In conversations with Harold McGee and through my own research, I found that most blue cheese, like Roquefort, Gorgonzola, and Stilton, contain butyric and caproic acids, among other compounds, which contribute to their *gooroot*-like, piquant flavor. I've also heard that in Lebanon, a similar product to *gooroot* is referred to as Shami (Lebanese) Roquefort. Considering the similarity in taste between *gooroot* and blue cheese, it is an example of cultural bias that some in the East have an aversion to "Western moldy cheese" but enjoy *gooroot*. Either way, we can say with funky flavored foods, like the advertisement for Marmite, "You Either Love It or Hate It."

I had heard that the animal funk in *gooroot* may have come from what was referred to euphemistically as "accidental contamination"—that is, as a result of urine and droppings, which might dissolve unnoticed into the warm milk during milking. I can see why that theory exists, but I could not find any evidence to support it. No doubt it might have happened occasionally, but as described before, I have seen pots with narrow openings in Āzarbāijāni villages for collecting milk, which reduces the chance of contamination. Today, the use of mechanical milkers, which are attached to the teat, means that if contamination had been common in the past, there would be a noticeable flavor difference between the *gooroot* made then and now, but there is not. Phew!

Overall, there was limited research on *gooroot*; most documents were anecdotal. I found only a handful of useful references and a few scientific papers about its chemical composition and health benefits in English and Fārsi. Online, there are many posts by travelers from around the world who have eaten *chortān* or *gooroot* and share their experiences as well as food writers who give background and recipes. There is a wide array of names and spellings for *gooroot* in English, often based on how each author hears a word or how it was pronounced to them. There is no standard romanization/transliteration, so none of them is either right or wrong. This has led to a confusing collection of spellings, some using *q*, *g*, or *k*, resulting in *gooroot*, *qurut*, *kurt*, *gurut*, and other variants. I have joined the fray and added my version, *gooroot*.

Even within Turkic languages, there are many different names for *chortān* and *gooroot*, such as *kurut*, *kesk*, and *kishk*, to name a few. *Gooroot* is also referred to as *panir*, cheese, or dough, likely due to its similarity with other shaped, aged, and stored curds. When making mozzarella in Italy or *rushan*, a cheese from Yunnan in China, the curds are pulled or kneaded to bring them together; therefore, we can see why it might also be referred to as dough.

One thing is certain: *Chortān* and *gooroot* were well received by all who encountered them and were adapted to each region's culinary practices. As mentioned earlier, *gooroot* in Turkic languages means "dry it"; when it was introduced in other regions with different languages, it took on a similar descriptive name in that language. It is highly probable that *gooroot* was introduced to Iranian peoples by Turkic peoples. *Kashk* is similar to *khoshk*, which means "dry" in Iranian languages and is most probably where it gets its name.

I found more texts on *kashk* in English and Fārsi than on *gooroot*, both etymologically and in terms of recipes and scientific studies. In the English documents, some papers have suggested that the word *kashk* comes from a Fārsi word for *kashkak* (meat cooked with grains) or *kishk*, a *tarhana* of sorts (discussed briefly in chapter 7), and though well meaning, these are misleading and have caused some confusion. I was surprised that the Turkic word *gooroot* was not given more attention.

Gooroot and *chortān* were important foods, providing nourishment and well-being and aiding in the survival of the people of Central Asia. These were used plain or made into more elaborate dishes with the resources they had. As their understanding of medicine evolved, food was also used therapeutically. Traditional *gooroot* dishes almost always include mint and walnuts. In the Unani traditional medicine of the region, all food is considered medicinal, and meals are meant to be balanced using "hot" and "cold" ingredients. Interestingly, modern studies into the antimicrobial phytochemicals in the mint family suggest they help preserve dairy products and aid digestion, which adds to the complexity of why these ingredients are used together.[5] Whether by chance or intention, people have found ways to make dairy more bioavailable but also to pair it with ingredients that aid digestion and reduce contamination or toxicity.

An example of this balance was when I had fresh *gooroot* (known there as *lor*) in a fish restaurant in Bandar Abbas in the Persian Gulf. It was served with spiced Piarom dates, as both a starter and a dessert. I was intrigued. They explained that *gooroot* and dates are "hot foods," eaten alongside "cold foods" like fish (a staple in the Persian Gulf) to create a balanced meal. In Āzarbāijān our ewe's milk *gooroot* is "cold," and cow's milk *gooroot* is "hot." I suspect their *lor* was from cow's milk.

Traditional *gooroot* is low in fat and high in calcium and protein. *Chortān*, depending on what type of curds are used, has more fat and live bacteria, and its heat-sensitive micronutrients remain intact. Both have health claims; *gooroot* is said to give you physical strength and good teeth and bones among Central Asians. It was eaten by Iranian *pahlavans* (heroes)—*the* protein shake of its day.

Having said all this, we must take into account that too much salt in the diet reduces the body's ability to absorb calcium. In Kazakhstan, studies have examined salt levels in *qurt* and the consequent health implications, such as increased blood pressure and metabolic diseases.[6] Both *chortān* and *gooroot* should be eaten in moderation. Another note on salty *gooroot* is that when cooking with it, it is best to under season the dish and only adjust the seasoning after adding *gooroot*.

We always boil *gooroot* when cooking with it in Iran, either on its own or in the dish it is added to. Growing up, I do not remember seeing *gooroot* eaten raw, as it is in some countries in Central Asia. That said, several studies in Iran found that *gooroot* was one of the least contaminated fermented dairy foods. Still, the habit of boiling *gooroot* is one I have not broken, even though the *gooroot* I buy in the UK is pasteurized.

To make or not to make *gooroot* at home? For someone writing a book about fermented dairy, it seems odd to dissuade you from making your own *gooroot*. But we have to remember, *gooroot* was made to preserve a precious natural resource utilizing the low humidity of the arid steppe. That combination of these two factors is what gives *gooroot* its special taste. We as cooks and gourmands are after flavor.

Bee Wilson says in *The Secret of Cooking* that our predecessors had no choice; well, we do. Jenny Linford's book *The Missing Ingredient: The Curious Role of Time in Food and Flavour* highlights the importance of time in food. So maybe there is a case for outsourcing *gooroot* making.

Making *gooroot* in humid environments without access to good raw milk and a well-established unpasteurized yogurt might be fun, but it won't taste the same. It would be like offering someone homemade Roquefort, possibly delicious, but it would lack the flavor of cave-aged cheese. To get a *gooroot* hit without the real thing, it is better to use commercial *gooroot*. It is safe; though it lacks some of the nutrition, it'll bring animally funk to your plate.

Having said all this, I like making food from scratch. If you would still like to make fresh *chortān* and accept that it will not be animally or funky.

I am still experimenting, but I have found a few ways to bring something of *gooroot*'s funk to recipes using dairy-free substitutes as well, which I share at the end of this chapter.

Making *chortān* in Bath during April was quite an undertaking. I ended up using an oven overnight just to dehydrate a handful of balls, making the whole thing uneconomical for such a small quantity. Still, the reduction in weight and volume was impressive. From one kilogram (2.2 lb.) of yogurt, I ended up with fifteen light, walnut-sized balls.

The taste was like dried yogurt. There was a hint of goat, but overall, it was clean, sour, salty, and mild in taste. I did not want to store them at room temperature, so I then froze them; freezing them meant the flavor would not develop any further, and it wouldn't taste funky.

I used them grated finely over dishes where I would have added *gooroot*, yogurt, or cheese. *Chortān* isn't hard enough to require a *seyin* and *gooroot ali* to rehydrate them.

To make a dairy-free *chortān*, follow the instructions below using thick, plant-based yogurt.

When making dairy-free dishes where I need the flavor of *gooroot*, I mix plant-based milk into plant-based blue cheese, which I buy. This gives the same look and brings some of the flavor. If you use these plant-based ingredients in a dish, do not call the dish *chortān*, *gooroot*, or *kashk*, because it is not. I call dishes such as *kashke bādemjān* traditionally made with aubergines and *gooroot* "aubergine delight," instead.

Chortān is a preserved food; therefore, you need to start with clean equipment and containers. With all traditional, and especially fermented, foods, one cannot be too prescriptive. Some of you will be making *chortān* in hot, humid climates; others in cool, dry ones. You may have pets, live in an air-conditioned apartment, or live in a home where the windows and doors are left open all day. Each environment, along with its unique microbial life, will shape the final product. You will have to observe, taste, and judge its progress as you would with sourdough, sauerkraut, or other live food. Each batch you make will be slightly different, even if you are using the same brand of plain yogurt.

Ideally, I would have used a full-fat, flavorsome ewe's or nanny's milk yogurt. I could only get pasteurized ewe and nanny yogurt (labeled sheep and goat).

Chortān

If making *chortān* in average temperatures above 19°C (66°F), then reduce the number of days listed below by half. For example, where I say to leave it at room temperature for six days, in warmer temperatures, reduce it to three days.

Ingredients

- 1 kg (2.2 lb.) of full-fat yogurt—the best you can get (if it is "Greek-style," that is already strained yogurt, and the procedure below will take less time)
- 1 tsp. of finely powdered natural salt

Leave the yogurt (if shop-bought, unopened) at room temperature for six days—where the temperature might be 19°C (66°F) or less.

After a week, open the yogurt. It should smell sour without any mold growth. If there is mold, you can either collect the mold and throw it away or start again.

Provided the yogurt is OK, pour it into a cloth-lined sieve. Cover it to keep debris from falling in, and leave it to strain for another day at room temperature.

By this time, it will have soured and the volume reduced. Add the salt, and mix it thoroughly; taste and adjust to your liking.

The *suzma* will now be stiff enough to roll into balls. Take walnut-size bits of the yogurt and roll them into small balls.

Put the balls on a parchment-lined oven tray and dry them out in the oven or a warming drawer at a steady temperature below 45°C (113°F), either overnight or for eight hours.

When you take them out, you will notice they are much lighter and roll easily around the tray.

Crack one open to be sure it has dried through to the center.

> ### Dairy-Free "*Chortān*"
>
> To make it, strain thick, plant-based yogurt in a cloth-lined sieve. Mash in some shop-bought, plant-based blue cheese with a fork. I usually don't dry it and use it fresh as a *gooroot* substitute in dishes.

Making *suzma* or straining yogurt to make *chortān*

This chapter began with the translation of the first lines of Raisa Golubeva's poem "A Precious Stone." Golubeva's poem was inspired by Gertruda Platais's memoirs. Gertruda was German and was imprisoned in Akmolinsk Camp in Kazakhstan for the "Wives of Traitors to the Motherland" in the twentieth century. She recalled when she and other prisoners were returning after collecting reeds and local Kazakhs started throwing rocks at them. At first, they were scared, as they were also being taunted by their guards, who implied that everyone hated them. When one of the women fell to the ground, her face came near one of these rocks, and it smelled like milk. The prisoners collected the rocks and took them

Chortān balls

back to the prison, where Kazakh inmates told them they were *gooroot*. The prisoners realized the locals were helping them out. In a song also inspired by this story, Galym Žajlybay says, "These are the most prized stones on earth / For they have the most heavenly taste."

In Kazakhstan and across many former Soviet republics, *gooroot* is associated with resilience. *Gooroot* reminded the Uzbek hero Amir Temur, also known as Tamerlane, founder of the Teymurid Empire in the fourteenth century, of his mother. It was part of Changiz Khan's army's rations, either eaten in the saddle or rustled up into a meal. This allowed for light travel and good-quality nutrition on the hoof, all while his forces brought great terror and death to my part of the world. Later, Ottoman soldiers had it in their rations. During the recent pandemic, *gooroot* and *kumis* were sought out by Kazakhs for their nutritional properties and, I suspect, for a sense of safety in the familiar. For us in the diaspora, *gooroot* is another taste of home, a source of identity, and a cause for bonding when we meet others who also eat it.

7

A Sour Sidekick

Talk about squeezing out every last drop: The final drops of milk are reduced to *gooroot*'s sour sidekick, *gara gooroot*, or *black gooroot*, which is made from the leftover whey from *gooroot* making. *Gara gooroot* is very sour and salty, mainly used as a flavor enhancer. Rather like the use of lemons or vinegar at the end of cooking to brighten a dish, *gara gooroot* is used sparingly as a souring agent when those ingredients are not available.

Any type of whey can be used to make *gara gooroot*. *Gara gooroot* may be prepared using either a single type of whey or a combination of different whey. The whey is boiled until it becomes dark and thick, similar in color and viscosity to treacle (molasses), and this is where it gets its name. In the past, hot rocks would have been used to evaporate the water, a method employed by many cultures around the world. I came across it in Canada when I visited a maple syrup farm, where we were told the same technique was used by indigenous people to make syrup.

Alternatively, this can be achieved by sun-drying the whey in hot and arid places like Central Asia. For this, the whey is poured into wide, shallow, metallic trays covered with cloth to protect it from dust and debris and left in the sun to reduce. Sun-dried *gara gooroot* is said to be finer, tastier, and lighter in color.

The very sour dehydrated paste is scraped into jars and stored in a cool, dry place. I have even seen it poured into pretty molds with a sprinkling of nigella seeds (in the bottom). Properly stored, it keeps for months, if not years.

Raviyeh Khanum, who lived in the village of Sār, used to make *gooroot* for my family using her own cow's milk. She told me that her sister ate a lot of *gara gooroot* for bone health. To make it in the city (Tehran), she would buy ten kilograms of yogurt, strain it, and use the whey, which they called *joy su*, or "green water."

Outside Iran, I found a few references to *gara gooroot*. Even where it is known, the name typically retains its Turkic origin. In Iran, it should be called *kashke siāh*, or "black *kashk*," but it is not. I've also come across the word *tarf* in Iran for a similar product.

Outside the region, the only other similar product I have encountered is made by the Dehong Dai people of Yunnan Province in China, who boil down the liquid left from fermenting radish leaves to make a thick, sour paste used as a dip.

Some cultures love sour tastes. If you have Iranian friends, you will know of our love of all things sour, so much so that we cannot wait for plums to ripen but eat them in their rock-hard, tart, green stage, our mouths puckering. It is no surprise that sour *gara gooroot* is also a snack. In Iran, you can buy little *gara gooroot* bonbons, some plain and others flavored with fruit.

One of the more bizarre applications for *gara gooroot* was recounted in food historian Charles Perry's story, where a car radiator was patched up with very thick *gara gooroot*.[1]

Gara Gooroot

You need two ingredients to make *gara gooroot*: a lot of whey and patience. Depending on what type of whey you start with and how long you boil it for, each batch will result in a different color and level of saltiness and sourness.

Ingredients

2 L (4 pt.) of whey (this can be one type of whey or a mix of different whey)
1½ tbsp. of salt

Pour the whey and salt into a shallow pan.
Bring it up to a boil, then turn it down to a medium-high heat. Stir now and then until it reduces by two-thirds.

> After this time, you will need to stir it regularly to check that it is not catching or burning on the bottom. It will continue to reduce and, in doing so, get thicker and darker.
>
> It is ready when you can draw a line with your spoon along the bottom of the pan and the paste remains on each side of this line. Scrape it into a small jar. Leave it cool before sealing and storing it at room temperature.
>
> A modern trick is to slake corn flour with a couple tablespoons of cold whey and add it to the whey when it comes to a boil. In this method, you need to stir constantly, and to my mind, this defeats the object of reducing and intensifying the flavor, but it does make it thicken quickly. It is up to you whether you decide to do this.

I have not found *gara gooroot* for sale in the West, and it is hard to find online too.

I will briefly talk about *tarhana*, which is not fermented dairy per se but made from a blend of fermented dairy like yogurt and grains such as flour, which is then left to ferment. In Turkey, a special "*tarhana* herb," called *Hippomarathrum cristatum* or *Echinophora sibthorpiana*, is added. This helps reduce potential pathogens during fermentation and adds to the flavor.

Making *tarhana* is an elaborate process that can take days or weeks depending on the climate. *Tarhana* is usually cooked in water with the addition of some fresh herbs or vegetables, turning it into a nourishing soup. It is generally called *tarhana* throughout, with various regional accents from Iran to Greece, and is known as *kishk* in Arabic.

The origin of the word *tarhana* may be from the Fārsi words *tar*, meaning "wet," and *khana*, meaning "food," "eating place," or "soup." I have included *tarhana* here because it often appears in Western references to *gooroot*. This may well have stemmed from the misunderstanding of the etymology of the word for *kashk* as presented in chapter 6.

8

Spinning, Laughing, Dancing to Her Favorite Song

The fat content in milk plays a pivotal role in the texture and flavor of dairy products. Fat not only enhances flavor but also lingers on the palate, contributing to a luxurious mouthfeel. Eating and drinking are multisensory experiences that engage the senses of smell, touch, and taste, creating enjoyment that is often irresistible. Our brains, which receive and interpret sensory information, are composed largely of fat, requiring good dietary fat for optimal function and well-being. Furthermore, consuming fatty foods activates the brain's reward systems, inducing feelings of satisfaction and pleasure. Therefore, the type, quality, and quantity of fat are crucial. The higher the fat content in the milk, the better the result of the products discussed in this chapter.

Traditional foods such as eggs, meat, and milk historically contained more beneficial fatty acids than many modern equivalents. Mass-produced eggs, chickens, and milk lack the micro- and macronutrient diversity found in products from healthy animals raised in natural environments, as previously explained.

We have seen how in unprocessed, natural milk, cream naturally rises to the top. This occurs because the large fat molecules present in the milk have a lower density than the surrounding liquid, causing them to float up and aggregate into a distinct layer of cream. This phenomenon is reflected in the Fārsi word for cream, *sarshir*, which means "the head of the milk." The English idiom "cream rises to the top," which is derived from "cream of the crop," metaphorically suggests that the best ideas and people will eventually stand out. In both cases, "cream" symbolizes excellence.

When the thick cream on top of milk is collected, it is typically consumed soon after or stored for a short while in a cold place, as it doesn't keep for too long. In Central Asia, cream is consumed fresh and also used to make several products, but the most popular are butter, ghee (clarified butter), and *gaymākh* (thick cream).

When sufficiently whipped, cream yields butter and whey, also known as buttermilk. In the following chapter, I also briefly discuss butter made with cheese whey, unsurprisingly called "whey butter." The type of milk, the season, and the type of food the animal has consumed influence the fat content and, consequently, the color and flavor of the butter. The resulting butter is used to enrich tea and for cooking and is also eaten as is. Butter acts as both a flavor enhancer and a carrier, emulsifying sauces and making them glossy. It can be used plain, whipped, smoked, caramelized, melted, or cold. Butter can also be flavored with sweet, savory, spicy, and luxury ingredients like truffle or champagne, to name a few. These types of butter are known as compound butter.

Depending on the culture and era, it has been considered food for either the poor and country folk or the wealthy. Though highly valued and seen as a special treat for herders, butter is used less in hot countries than in cooler ones. In places where oil-producing plants like flax, sesame, or olives were abundant, vegetable oils were preferred, and dairy products were looked down upon. So much so that in the first century AD, Pliny the Elder regarded butter as the food of "barbarians" from the empire's northern regions—where they had access to milk. He even went on to say that butter was fit only for topical use to soothe burns rather than as food. However, in Central Asia, butter was considered a great delicacy, consumed fresh, reserved for special occasions, and used in rituals.

As you know, I enjoy looking into etymology for clues about the history of products. The English word *butter* is said to come from Greek into Latin and then English, ultimately meaning "cow cheese." The color of cow's milk butter influenced other nouns, such as *butterfly*. In some accounts, this is so named because butterflies appeared during butter-making season; in others, brimstone or clouded yellow butterflies were said to be the color of butter, and they flew, hence Papilionoidea go by the common name "butterfly." Similar lore exists for buttercups, a bright yellow flower. Growing up in the UK, my sister, our friends, and I would hold buttercups under our chins, and if there was a yellow glow on our skin, it was said to mean we liked butter.[1]

Moreover, the English language has many butter-related expressions, such as "bread and butter" (to allude to the main part of someone's income), "butter someone up" (to flatter), or "know on which side your bread is buttered" (to understand and act in one's own advantage). In contrast, the lack of butter-related idioms in Central Asia might suggest that butter doesn't feature as prominently in these cultures.

In Turkic languages, the word *yāgh* refers generically to all fats and oils, and butter is called *kareh* or *maskeh* in Turkic and Iranian languages of the region. Sometimes, the word new or fresh (*tāzā*) is added before *yāgh* to distinguish butter from other fats like tallow.

In general, there are two types of butter: sweet and cultured butter. Sweet butter is made from fresh cream, while cultured butter is produced using yogurt or fermented cream. The by-product is buttermilk, which is whey, similarly categorized as sweet or acidic. Butter consists of milk fat, water, and, if added, salt.

To extract butter from the cream or yogurt, it needs to be churned. In Turkic languages, this process is called *chālmākh*, meaning "beaten," while in Iranian languages, it is expressed as *kareh ro az māst migirand*, meaning "they catch butter from yogurt." Whether by beating or chasing, the agitation disrupts the fat globules. As these globules bump into one another, they stick together and form clumps of butter within the buttermilk. Butter can be churned by hand or using churns made of skin, pottery, or wood. There are various manual churns used around the world, including both horizontal and vertical designs. Horizontal churns may use a waterproof animal-skin sack that is shaken back and forth, while vertical churns employ a plunging motion. Settled cultures tend to use heavier or more elaborate churns, while nomadic groups use simpler, more portable versions.

In many cultures, making butter is traditionally a woman's role. Butter making is best in cooler months, as it can become messy in hot weather. Keeping the butter-making tools cold, such as hands, churn, and bowls, is also recommended. Churning is a laborious task, and some women sing while they churn. In some cases, two people, one at each end of a suspended churn, work together and sing in tune with the rhythm of their churning. I listen to Nora Jones daily, and her lyrics used in the title of this chapter made me think of the women churning butter.

Churns are known as *nehra* or *yāyik* in Turkic and *mashk* in Iranian languages. When visiting the village of Sār in Āzarbāijān recently, I saw butter made in a *Nehra*; these are earthenware, amphora-like pots with a narrow neck and a small hole near the top. It is placed on a cushion and rocked back and forth. The yogurt inside swishes, and then eventually you can hear thuds of butter inside. The small hole is poked open, and a piece of straw or a skewer is inserted to feel for clumps of butter. If the butter has formed, the *Nehra* is emptied, the whey is collected, and the butter clumps are taken out, washed in very cold water, and then squashed by hand to form balls of butter with characteristic finger marks.

Nehra

Similarly, when I visited nomads in Iran in 2014, I saw their simple churns assembled only when needed. These consisted of a large skin bag shaped like a bolster, suspended from a tripod made of wooden poles. Freshly made yogurt was poured into the skin, which was tightly fastened with rope, and then the skin bag was rocked back and forth. The swishing sound changed as the butter began to form in the buttermilk, and you could feel it knocking around inside. Toward the end, they added ice-cold water to the bag and continued rocking it. The contents were then poured out, and the butter was separated and kneaded in another bowl. Some of the buttermilk was reserved. Cold water was poured over the butter repeatedly until the water ran clear. Finally, the butter was shaped into large balls by repeatedly tossing it from hand to hand, allowing water to drip out.

Making butter by hand is the most labor-intensive method of churning. Fresh or soured cream is kneaded for at least an hour until the fat separates into butter and buttermilk. This process demands patience and strength, much like journalist Michael Pollan's sentiment: He once said that if you want something to eat, make it from scratch. First, you will know its worth; second, you will expend energy countering some of the potential weight gain from consumption; third, you will learn patience; and finally, you won't waste a drop.

Butter is typically made in small quantities and consumed within ten days. Before refrigeration, butter was stored in a bowl of cold water placed in a cool spot, with the water changed regularly. Butter can go rancid quickly, as it contains water, which fosters bacterial growth. Although rancid butter is not harmful, it has an unpleasant taste and aroma. This same moisture in butter is also the reason why it hisses or spits when heated. Like any fat, it should be stored in a cool, dark place, such as under a bench in the coldest part of the yurt. In Āzari, your land or country is your *yurd*, which is the origin of the term. We also call nomadic tents *obba*; Mongolians call theirs *ger*. Talking of Mongolians, they have *suutei tsai* (milky tea), known in the West as butter tea. This is a savory drink, usually made with green or black tea leaves, salt, milk, and butter whisked in. It is a warming and high-calorie drink necessary in harsh conditions and an acquired taste. Butter storage depends on where you live. In Ireland, butter has even been found preserved in bogs.[2] Butter lasts longer in a freezer, but it can crystallize and develop a grainy texture.

Historically, we mainly had cultured butter made from ewe's milk in Iran. The butter is light cream or white and, as described above, comes in large, knobbly, grapefruit-sized balls or in individual hand-shaped portions, the finger indents forming delicate, leaf-like patterns revealing its handmade preparation. Nowadays, this type of butter is usually served with *chillo chabāb* (rice and meat on a skewer) in traditional *chillo chabābi* restaurants in Tabriz. Today, industrial yellow butter, packaged in aluminum foil, is widely used. Traditional butter is white because it is made from ewe or nanny's milk.

Have you ever wondered why milk is white and butter is yellow? Well, it is because cows process the beta-carotene (a yellow-colored pigment) present in fresh grass and wildflowers and store it as body fat. Although their milk is white, the pigment is in the fat within their milk. Churning milk into butter alters the fat molecules in milk, which releases the pigment and results in yellow butter and whitish buttermilk. Many traditional dairy products still eaten in Central Asia are mainly made from ewe, nanny, and buffalo cow's milk and are white in color. This is because these animals do not store the pigment in their fat but instead convert it into vitamin A, which is colorless.

In an ideal world, we could all buy unpasteurized butter made from the milk of animals that graze on seasonal pastures, like those from France with a protected designation of origin (PDO). To be sure dairy products are carefully made, we must look out for information on packaging. If all butter and dairy were produced this way, there would be no need for special labeling.

This is yet another reminder of the importance of animal welfare and healthy pastures, which leads to more flavorsome and nutritious milk. Feeding animals grains or fermented silage is ultimately harmful to the animals, the environment, and to us. Industrially produced butter often contains coloring and salt. It is usually stored and or transported around the world, which means it is no longer fresh. Interestingly, butter in the UK is now cheaper than it was at the beginning of the twentieth century, but so is the quality. As the saying goes, "You do not get anything for free."

For those who are unable or choose not to consume dairy, there are commercial plant-based nondairy hard blocks that can be used; most, but not all, are full of synthetic chemicals. Alternatively, you can search online and experiment to make your own plant-based fats if you need them. I agree with the late chef Anthony Bourdain, who once said, "Margarine? That's not food. I Can't Believe It's Not Butter? I can. If you are planning on using margarine in anything, you can stop reading now, because I won't be able to help you."[3]

Sweet Butter

To make sweet butter, start with the richest, creamiest cream you can find; this is essential. The next thing is to make it on a cold day and with cold equipment and ingredients.

Ingredients

1 kg (2.2 lb.) of fresh cream

A few hours before making the butter, chill the bowl you plan to whip the butter in. Sometimes I'll put the beaters of my whisk into the fridge as well.

Pour the chilled cream into the cold bowl and set your mixer on medium to begin with.

Cream in a bowl

After five minutes of beating on medium, turn it up to the highest level and beat the butter for another fifteen to twenty minutes. As you whip, the cream will change color (if you are using cow's milk cream, from white to yellow). Eventually, the buttermilk will start to separate, leaving you with clumps of butter and watery buttermilk.

Overwhipped cream

Butter and buttermilk

At this point, remove the clumps of butter from the bowl, and knead them on a board, ideally on a sloped surface to help the buttermilk drain out. I create a slope by rolling a tea towel under the wooden chopping board I use to knead the butter.

A lump of butter

After pressing out as much buttermilk as possible, wash the butter in ice-cold water. To do this, return the butter to the bowl and knead it in the cold water. You may need to change the water a few times until it is clear.

To remove excess water from the butter, grab handfuls of butter and gently bounce them in your hand over the sink or a cloth. At this stage, you can season it with fine salt or portion it and flavor it if you wish.

To shape the butter into a log, put it on greaseproof paper, then roll it in the paper and twist the ends of the paper. Then roll the whole thing until it is a uniform log. You can also buy wooden butter paddles for shaping butter.

Butter wrapped in greaseproof paper

Slice off what you need for immediate use, and place it into a butter dish and store the rest in the refrigerator. It will keep in the fridge for a month.

Cultured Butter

The process of making cultured butter is the same as making sweet butter, except here you begin with fermented cream or yogurt. It is up to you whether you would like to sour the cream by leaving it to ferment naturally or add a starter such as kefir before proceeding to make butter.

Once your dairy is fermented, follow the same steps as for sweet butter: whip, knead, wash, and season.

Another idea would be to split the fresh cream, make butter with half, and then sour the other half to make cultured butter. That way you can try both types to see which you prefer.

As for the buttermilk or leftover whey from butter making mentioned earlier, it is watery, unlike the thick, creamy commercially available buttermilk. I discuss it in the final chapter of the book, which is dedicated to whey.

Now we'll look at the most traditional cooking fats in Central Asia. The first is animal body fat or *jān yāghi*, such as tallow (less so lard due to religious considerations), which has been and continues to be used as food and for therapeutic purposes. Today, in the modern world, we have access to a wide variety of cooking oils, many of which are highly processed and potentially harmful to our health. Historically, people used animal fats for cooking much the way we now use butter or cooking oils.

Interestingly, good-quality animal fat is one of the best options for high-temperature cooking. When consumed in moderation, it is often healthier than many vegetable oils. It contains fat-soluble vitamins A, D, E, and K. Fat-tailed sheep tallow, for example, is also rich in omega-3, omega-6, and omega-9 fatty acids. Animal fat is shelf stable but can turn rancid if exposed to heat and light. Stored in the fridge, it can last for years. Different animal fats vary in taste, texture, and color. They usually solidify at room temperature, creating a distinct mouthfeel, which may be perceived as cloying when eaten cold.

After *jān yāghi*, ghee or clarified butter is the second most popular cooking oil of the region. Ghee, as we call it in the West, is known as *sāri yāgh* or *sāri māi*, literally "yellow fat" in Turkic languages, and *roghan zard* (yellow oil) or *roghan heyvani* (animal oil) in Iranian languages. This too has a high smoke point and somewhat nutty flavor. Ghee is used for frying and dressing dishes, especially *pilow* (rice). Consumed in small amounts together with an active lifestyle, it is a healthy ingredient, particularly if you do not have any underlying metabolic health issues. Many regional stories and lore praise ghee for providing strength and vitality. It is even used medicinally, both internally and topically.

In most hot countries, butter is clarified to extend its shelf life. With its high fat content, ghee has been enjoyed for millennia as a food and ingredient. *Sāri yāgh* or ghee is essentially pure butterfat with its water and milk solids removed. In turning butter into ghee, we are removing the water so that there is less chance for bacteria to take hold and ruin it. Though time-consuming, it is worth making, as you will end up with a shelf-stable product with a wonderful aroma and golden color. It solidifies below 15°C (59°F) and can be stored for several months in a cool, dark place without refrigeration.

You can make *sāri yāgh* at home using high-quality butter. If you would like to try an alternative to cow's milk ghee, consider sourcing ewe or nanny milk cultured butter.

Before starting, ensure all jars, lids, jugs, ladles or spoons, and anything else that is going to come into contact with the ghee are clean and dry. To do this, I wash everything, including the cloth, and leave it to air-dry.

I put the equipment on a tray in the oven at 110°C (230°F) for ten minutes before the end of the ghee-making time. Remove these five minutes before the ghee is ready, so the utensils are still warm and sterile, but not too hot.

Every batch is different, and the timings will vary, so you will need to watch it carefully. You need to evaporate the water without scorching the bottom. It is a delicate balance that requires attention. With practice, you will develop a sense for it. The bubbles on the surface are very informative; their size and the sound they make when popping can tell you a lot.

Sāri Yāgh

Ingredients

2 kg (4.4 lb.) of good-quality butter

Start by gently melting the butter in a heavy-bottomed pan, taking care not to let it burn. Keep it on a steady heat, and adjust the heat as needed to achieve this.

As the butter melts, it will turn opaque, and a fine foam of tiny bubbles will start to appear on the surface. Turn the heat up to medium-low.

When the melted butter heats up, the bubbles will change, and a foam will gather at the edges of the pan. Skim and discard this foam.

Then the bubbles in the foam will start to get larger, while the milk solids, starting to caramelize, begin sinking to the bottom of the pan.

Your goal is to evaporate the water without burning the butter or disturbing the sediment. In order to do this, adjust the temperature accordingly and stir gently only if needed.

After about twenty minutes, the melted butter will start to become slightly darker. When it is clear and golden, remove it from the heat and strain carefully through a cloth to catch the sediment.

There will be caramelized milk solids in the bottom of the pot; this is known as *tortā* in Turkic. This is a cook's treat: crispy, flavorful little drops of fat that can be eaten as they are or used as an ingredient.

The final Central Asian creamy delicacy I highlight is *gaymākh* (Āzari), also called *sarshir* in Fārsi and *khāme* or *kerem* regionally. Thick and sticky, it is prized for its rich texture and distinctive flavor.

I've explained before how cream naturally rises to the top in whole milk; simultaneously, proteins in the milk, when exposed to air, denature and coagulate, forming a skin on the surface, or lactoderm. Repeated heating of milk thickens this layer of cream and protein, which is how we get *gaymākh*. The proportions of cream and protein in *gaymākh* vary depending on the milk's composition, temperature and duration of heating, and the number of times the process is repeated.

Gaymākh

Ingredients

2 kg (4.4 lb.) of whole buffalo milk

Gently heat the milk to around 80°C (176°F) for approximately ten minutes, stirring occasionally to prevent scorching at the bottom.

Pour the hot milk into shallow trays and refrigerate overnight.

Once cool, loosen the edge of the crinkly cream from the pan with a knife, then collect it.

Depending on the milk's fat content, this process can be repeated multiple times to get thicker *gaymākh*.

Enjoy the *gaymākh* fresh or refrigerated, and use within a few days.

You can also make a lightly fermented version: To do this, leave the raw milk at room temperature overnight, then follow the procedure above. This results in a subtly tangy *gaymākh*.

I use the remaining milk to make yogurt.

Gaymākh is a popular breakfast food eaten with honey (known in Turkic languages as *bāl gaymākh*) and a perfectly brewed *chāi* straight from the top of the *samovar*, a traditional water heating urn. It also pairs beautifully with fruit preserves or baklava.

In the UK, clotted cream, also called Devonshire or Cornish cream, is eaten similarly with jam.[4] Traditionally served with scones and jam, there is debate as to whether the cream or jam goes on the scone first. I prefer cream first and jam on top—a choice perhaps inspired by the way we eat *bāl gaymākh*.

Morish, high-fat, high-sugar foods like cream and jam on scones or *bāl gaymākh* are simple luxuries; such combinations of fat and sugars do not exist in nature. Consumed occasionally (as a treat) and in moderation alongside physical activity, they pose little harm due to their use of natural, simple ingredients. Industrial food manufacturers have exploited our innate delight in such combinations, producing ultraprocessed foods that mimic these traditional indulgences but lack nutritional value. Since the 1960s, the food industry has identified and targeted our "bliss point"—the optimal combination of sugar, salt, and fat—to create addictive, irresistible products. Containing synthetic fats and sugars and heavily marketed through persuasive packaging, they have wreaked havoc on public health on a global scale. These "foods" are addictive and have been developed further since the 1970s by some tobacco companies that bought out food companies.

Gaymākh rolls

I'll end this chapter with two butter-related personal anecdotes. The first, a constant in all my kitchens ever since it was gifted to me by my Hossain Dāi (maternal uncle, mother's brother), is a handmade butter dish. It is one of the first things I unpack when I move. Somehow, this now-chipped dish makes the place instantly feel like my home and kitchen.

The second is about my Auntie Nesta and her love of butter. In her North Wales kitchen, she had a stainless-steel bread bin and a stainless-steel butter dish, which she would put in the oven to soften the butter. Once softened, she slathered it thickly on hot toast for breakfast every morning and then again at teatime on little triangular cucumber or Marmite sandwiches with the crusts cut off. Every year, in late spring, when I see the first Jersey Royal potatoes of the season, I remember her gleefully eating them, drenched in a melted pool of Lurpak, her brand of choice.

I cook with butter regularly, and if I have leftover cream, I always turn it into butter, flavoring it with whatever is available in the garden or the spice drawer. I'm fortunate to have a fantastic farmers' market, the first in the UK, established by my friend Peter Andrews, where I can buy unpasteurized butter, whey butter, and artisanal ghee directly from the producers.

Homemade butter in a butter dish

9

A Separation: *Su & Dan*

In Turkic languages, we find two basic terms that speak to the antiquity of dairy separation techniques: *dan*, referring to the solid food or curd, and *su*, referring to water or whey. The remarkable simplicity and elemental nature of these terms, both being basic words, may suggest some of the earliest linguistic codification of dairy-processing knowledge in the region. During conversations with dairy producers, I found that these terms are still commonly used in contemporary dairy production. Therefore, I decided to include them in the title of this chapter, where we focus on *panir* (cheese) made from *dan* (curds).

Cheese is an essential part of the Central Asian diet and, in some areas, is eaten at most meals. In Āzari, instead of the phrase "bread and butter," we say *panir chorayh*, or "cheese and bread." We eat it at breakfast, at lunch, as a snack, or at the end of a meal.

We even have sayings about how and what to eat cheese with. I remember being told that if you ate bread and cheese without walnuts, *chorafahm olasan* (you will become dim). *Maghz panir*, or ground walnuts and cheese, was what we ate for breakfast. We now know that salty food can impair cognition and that walnuts are good brain food, especially when eaten first thing in the morning. Another example of modern science catching up with indigenous wisdom.

In the past, there would always be bread and cheese on the *sufra* at every meal, and one small *ticha*, a "little portion," was eaten at the end of the meal, rather like a cheese course in the West. Maybe this is where the custom originated?[1] Here, too, scientific findings show that eating cheese after a meal may help reduce acid erosion of tooth enamel.

In one of the oldest northern European languages, Welsh, the word for cheese is *caws*, which is similar to other European words derived from the Latin *caseum*, meaning "cheese." The term evolved into *caseum formaticum*, or "formed cheese," which led to the French *fromage* and Italian *formaggio*. In Fārsi, *cheez* means "thing," and this word is thought to have been adopted by British colonials in India, evolving into phrases like "Say cheese" in photography and "the big cheese."

While talking about names, the word *feta* means "slab" or "slice" in Latin. It was adopted as the name for Greek cheese after the seventeenth century; prior to that, cheese in Greece was called *tyrí*. In Homer's *Odyssey*, Ulysses encounters Polyphemus the Cyclops making cheese in his cave; this is said to be the first written account of cheese making in Western literature.

Interestingly, in both Turkic and Iranian languages, in many places cheese is called *panir*, with slight variations in pronunciation, except in Kazakh, where it is known as *irimşik*, and in Uyghur and Uzbek, where it is called *pishlaq*.

There are many stories about the origin of cheese, including one where nomadic tribes in Central Asia carried milk in animal-skin bags, which led to its chance discovery. The movement of the carrier, be it human or horse, agitated the contents like a churn, resulting in cheese. Another version describes bags of milk hanging by the yurt doorway, shaken by those entering and exiting for luck, leading to the formation of cheese. Looking at this more carefully, if the milk was stored in a bag made from an unweaned animal's stomach (containing chymosin) and the movement generated enough heat to activate the enzymes, then yes, the curds would be what we now call cheese. On the other hand, if the milk was stored in a hide bag kept in the cold, the agitation might have produced butter. We know that stored milk will clabber, and the curds from this, called *lor* in Āzari, are not yet cheese as I have defined it in this book.

Regardless, soon after, people likely realized that milk stored in containers made from the stomachs of young dairy ruminants produced a different kind of curd than that formed through spontaneous fermentation or backslopping. Could they have subsequently begun deliberately using such containers for this purpose? The development of containers and sieves facilitated the separation of the curds and whey, possibly leading to new dairy products.

Archaeological findings suggest that dairying began around ten thousand years ago, and pastoralist communities began developing sophisticated dairy-processing techniques around eight thousand years ago, particularly in the Caucasus. This is supported by the discovery of beeswax-lined containers (which waterproofs them) used to separate ewe's milk curds from whey. These particular finds were ascertained through genetic analysis of protein and fat residue found on pottery sherds in the area. If the curds were formed by rennet, then we could assume cheese making likely began with pressing the curds inadvertently, creating what we now recognize as cheese. Unstable surplus milk could be transformed into food that could be stored and consumed throughout the year.

It is widely accepted that cheese making spread westward from Central Asia, with the Romans further developing the craft. People all over the world are deeply passionate about their cheese. In the dairy world, professionals stake their reputations on the quality of their cheese, while homemade cheese is a point of pride, especially in Central Asia, where it is often the domain of the lady of the yurt. Artisanal makers know their milk intimately; they instinctively understand when to add the appropriate amount of rennet, the resulting curd type, and how best to handle it. Cheese making is both an art and a science.

Recipes for rennet, the enzyme used to make cheese, and the amount to use are often closely guarded secrets. In the past, brides brought rennet to their marital homes as part of their dowries or learned its preparation from their in-laws. Either way, rennet would have been another precious dairy asset or worldly good.

I am reminded of a line from *Monty Python's Life of Brian*, which, though spoken in confusion, rings true: "Blessed are the cheese makers." Cheese was and remains an important food. In fact, just as I was writing this, twenty-two tons of cheddar were stolen in the UK; no wonder Parmesan is sometimes stored in banks! Beyond its financial and nutritional value, cheese also represents the people and places associated with it. In some cases, it even becomes a status symbol, as exemplified by expensive cheese like Pule, a Serbian jenny's milk cheese.

Among all dairy products, cheese represents the greatest diversity in forms, flavors, and techniques of milk transformation around the world. Cheese varieties worldwide are numerous, each uniquely shaped by the methods, conditions, milk types, and aging processes. For instance, French ewe's milk blue cheese may only be called Roquefort if it is made and aged in the Combalou caves of Roquefort-sur-Soulzon. That said, it is now possible to make any style of cheese anywhere; using mold cultures such as *Penicillium roqueforti*, you can even make a nondairy Roquefort. While many Central Asian cheeses lack the same protections, some are recognized by the Slow Food's Ark of Taste, and thankfully, others are being considered for recognition as well.

As with all the other fermented dairy products in this book, making simple traditional cheese typically requires only a few basic ingredients: raw milk, a coagulant, salt, time, and a place to store the cheese. Factors affecting dairy fermentation, including the environment, play a significant role in creating mature cheese, which also requires patience. Aging or ripening cheese in a dry, humid environment or in brine produces very different results.

Aged cheese is managed carefully to undergo further development by the addition of microorganisms and the passing of time, refining flavor and texture. The French term *affiner* refers to this refinement process, during which temperature, humidity, and microbial conditions are carefully controlled. Removing excess whey during aging (since water is a breeding ground for unwanted bacteria) improves preservation and flavor.

Cheese can become moldy during aging, which is carefully managed by cheese makers. While cheese itself is not mold, specific molds are intentionally added to develop varieties of cheese with enhanced tastes and textures. Molds on hard cheese like Parmesan or cheddar are usually harmless and can be trimmed away; mold on high-moisture cheese like feta or ricotta indicates spoilage, and they should be discarded. In Central Asia, there remains a deep reluctance to eat moldy foods, as expressed in the Turkic term *chuflanmish*, meaning both "moldy" and "spoiled." There is an aversion to anything moldy; moldy foods are generally seen as bad and not eaten intentionally.

Cheese can be made from whole or skimmed milk or a mixture of different milk types, such as ewe's and nanny's milk. Higher fat levels yield better taste and texture in cheese. When making cheese, the quality of both the milk and the rennet is crucial, particularly their ability to produce strong curds. Milk from buffalo cows, ewes, and nannies is suitable for *panir* making. Cow's milk, and especially camel cow's milk, tends to produce softer curds. When curds are soft, sometimes calcium chloride is added to strengthen them.

The cheese making process begins with separating the curds from the whey, either by spontaneous fermentation or with the addition of acids and/or rennet. In chapter 3 on fermentation, I discussed the molecular changes in milk that bring about this separation. Kefir, yogurt, whey, sauerkraut liquid, lemon juice, and vinegar are a few of the other things that may induce curdling in milk.

Rennet is the main "cheese" maker, especially the enzyme chymosin, a fortuitous gift from the stomach of an unweaned dairy animal. The first cheese as we know it was likely discovered when chymosin was accidentally added to milk and started curd formation.

In my research, I found a few "recipes" for rennet. I was told "a bit of this and that" or "whatever you have to hand" rather than a precise set of ingredients. I saw and smelled many jars of what looked like ditchwater with things floating in it, sometimes even an egg. They almost always contained a piece of a lamb's or calf's (or another unweaned ruminant's) abomasum, also known as the "fourth stomach" or "rennet bag." Some of the people I spoke with referred to it as *barsagh*, while others called it *gursagh* in Turkic and *sheerdan* in Fārsi. I was told that the abomasum was salted during preparation and sometimes even dried whole. A small bit was then used or, as they described, "brought back to life"—by rehydrating it—whenever cheese was made.

Most of these rennet mixtures also included vinegar and fragrant herbs. Artisanal rennet is not considered "safe," and cheese made with it is generally not sold to the public in Iran. However, you may be served this type of homemade cheese if you visit a family or arrange to acquire it privately from pastoral communities.

Today, industrial, genetically modified, and synthetic rennet are widely used and come with clear instructions for dosage and temperature, making them more reliable and consistent. The four main types available to cheese makers are animal rennet, microbial rennet, plant-based rennet, and fermentation-produced rennet. The curd-to-whey ratio depends on factors such as milk composition, moisture loss, curd cutting (the curds are cut up or raked to increase surface area), and rennet type.

Cheese falls into three broad categories: fresh, fermented, and aged. Fresh cheese is made immediately after whey separation. It has a mild, comforting flavor and texture and can be savory or sweet. Some are fried to create a caramelized version, such as *gizil irimshik* and *eezgii*. Commonly known as *lor* in Turkic and Iranian languages, fresh cheese is a staple of pastoral herders' diets. It is eaten soon after being made, as it doesn't store well or travel easily. It looks and feels like cottage cheese or ricotta. My mother said that when she was a child, her father would make *panir* in February for the coming year. She and her siblings would eat small bowls of freshly prepared *lor*, sprinkled with sugar, as a snack. As mentioned in chapter 6, we encountered *lor*, or fresh curds. In most areas of Āzarbāijān, curds from clabber, cheese, or *gooroot* are all called *lor*, and when salted, *shor*.

The region has one main type of cheese, generally referred to as *panir*, which is very similar to what is known as feta, Greek cheese, or salad cheese in the West. It is made with rennet and then brined. It is white, holey, crumbly, sour, and salty, usually made with either ewe's or nanny's milk or a mix of the two. It differs from Indian *paneer*, which is usually produced with an acid coagulant instead of rennet.

The finest cheese in Iran is from Āzarbāijān, known as Lighvān *pan-iri*, or *panir* from Lighvān. Shepherding and ewe's milk cheese production remain the main economic activities there. The predominant local breed of sheep is the fat-tailed sheep *gizil* breed. *Gizil* can mean "red" or "gold," referring to their dark fleece, which is said to absorb more goodness from the sun, improving milk and meat quality. You may recall stories of a certain Jason chasing after a certain "golden fleece."

On a personal note, I feel a deep connection to and pride in Lighvān. My maternal grandmother's *yaylāgh*, or summer home, was located in the cool hillsides of this place. Parts of the old house still stand, with their carved sheeplike lions guarding the entrance. It overlooks the mountains, summer pastures, and a valley through which the Lighvān river flows. The river is fed by meltwater springs from Sahand mountain, an extinct volcano. The water from this river was used for brining Lighvān *panir*.

The resulting whey, or *su*, from *panir* making is not discarded; it is used to brine the cheese or for other purposes, which I discuss in more detail in chapter 10. I now live in the southwest of the UK, near Cheddar, where the famous cheese of the same name is made. I was delighted to find that a few producers here, and in some other parts of the UK and France, still make whey butter from cheese production. The remaining cream in cheese whey is extracted using a centrifuge and then churned into whey butter. The small amounts of buttermilk leftover from this process are either given to animals as feed, put on the compost, or thrown away.

Panir needs time to develop in brine. Much like vegetable ferments, where vegetables are chopped, brined, and left to ferment as lactic acid

Whey butter and cheese

bacteria (LAB) work their magic, the same applies to *panir*. Brining, also known as pickling, is the most suitable method of cheese preservation in the arid climates of Central Asia. For example, when you buy feta in small slabs sealed in plastic, it usually deteriorates quickly after opening, developing a slimy surface. You can rinse it and still use the feta, but it lasts much longer if placed in a simple brine—two teaspoons of salt dissolved in two hundred milliliters of water, stored in a sealed container in the fridge. Mr. Abdi, our local dairy shop owner, always offers us some *su* (brine) when we buy cheese. It is usually made with a 1:2 ratio of salt to water, which he claims "lasts forever."

When fresh curds are immersed in concentrated brine, the high salt concentration causes osmotic stress in bacterial cells and further denatures proteins. These modifications lead to a more consolidated protein network. The characteristic "eyes" in *panir* result from gas produced by active bacteria becoming trapped in the curd and leaving holes as it escapes. Together with the milk, these microbes influence the texture, flavor, and eye formation of cheese while inhibiting the growth of undesirable bacteria.

When making *panir*, salt can be added to warmed milk, layered in between the curds when packing them in liquid, or added to the brining liquid before the curds are packed into containers. Temperature and brining duration are critical to the final product. Typically, *panir* is brined for three to four months, developing a complex flavor profile that enhances its organoleptic properties. When the cheese is ready, it is eaten and does not keep beyond ten months.

To make *panir*, a small amount of rennet is added to already curdled warm milk and left to work its magic. Curds form within one to three hours; this depends on many factors, including the cheese maker's intuition. Afterward, they are gently "combed" with a utensil or clawed by hand to break them into smaller pieces, which increases surface area and helps further rennet action. Then the mixture is poured through a tightly woven cheesecloth and left to drain, either suspended or placed on an open-weave basket to let the whey run off. The curds are then pressed under a weight to get rid of more whey. My grandfather had beautiful, large, flat, white river rocks that he kept in the garden and washed before using them as *panir* weights.

The blocks of curd are packed into containers filled with either the whey from the cheese-making process (with added salt) or fresh brine and fully submerged, then sealed and stored. After three or four months, they check to see if the *panir jalip*, which translates to *panir*, has come or arrived. Even today, you can buy artisanal *panir* in five-kilogram tin-lined boxes. These need to be opened with an old-fashioned can opener, and the edges can be dangerously sharp.

Another traditional cheese that is now disappearing is *chup paniri*, named after the earthenware container called *chup* in Turkic languages or *kuzeh* in Iranian languages. These heavy, brittle pots are mostly used by settled peoples. The cheese is preserved underground.

A Separation: *Su & Dan* 121

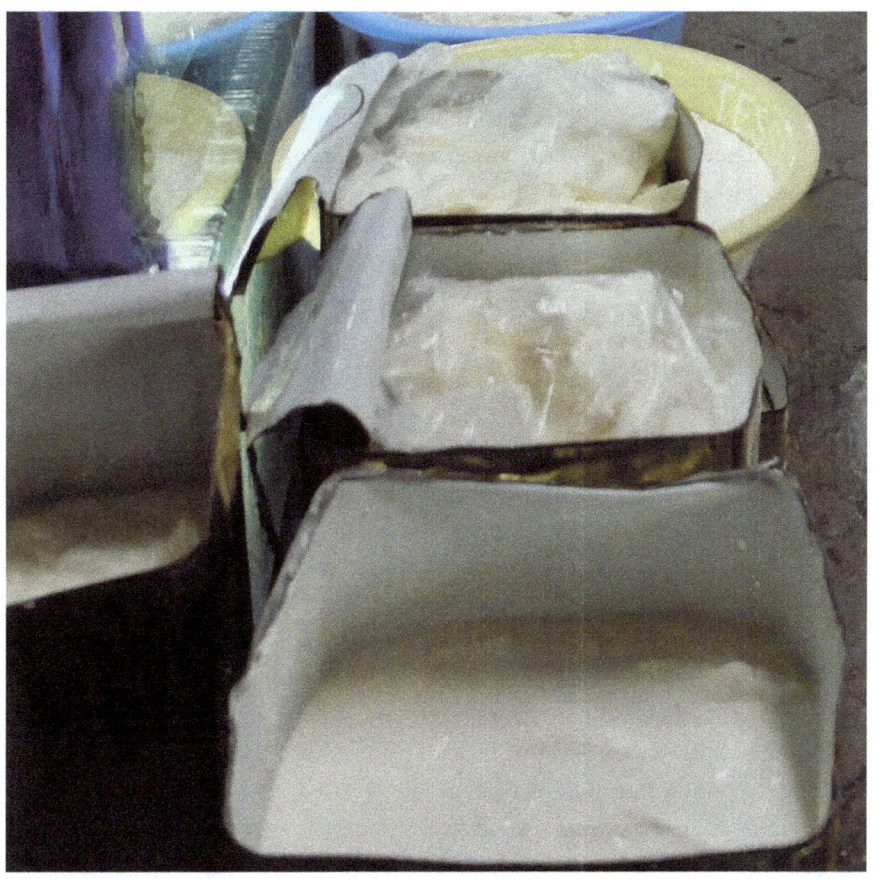

Panir in large tin cans

Chup paniri is made much like *panir*. After brining for about forty days, which is considered the right amount of time for salt to fully penetrate the cheese, the cheese is removed from the pot, broken by hand into small pieces, and sometimes mixed with aromatic herbs such as summer savory or thyme. It is then packed very tightly back into the pots and pressed down with something like a pestle to ensure there are no air pockets. Large, pliable grapevine leaves are placed on top, and the neck of the pot is tied with cloth.

The pots are then inverted and gently buried in a hole that has been lined with a layer of ash or sand. It is believed that the inverted position allows moisture to escape and air to circulate, helping prevent spoilage and enhancing nutritional value. The cheese is left to ripen for four to six months. When ready, the pot is unearthed, the leaves are removed, and the cheese is scooped out with a spoon, or the pot is gently cracked open and the cheese sliced to share.

If the cheese is ready but won't be eaten immediately, the old cloth is replaced with a clean one, and the pot's mouth is sealed with melted beeswax. It can then be stored for several more months. Nomads use a similar process but pack the cheese into animal-skin bags instead. These are sewn shut and stored in caves along migration routes or buried in soil to be retrieved on their return journey, by which time the cheese will be ready.

Similar cheese, collectively known as *chat paniri*, or village cheese, is made in many villages in the area. Sadly, every time I visit, fewer of these cheeses are available. Globally, small-scale cheese producers struggle to sell their products. Moreover, as young people show little interest in these skills, knowledge of rennet and *panir* production is at risk of being lost when current producers retire. Meanwhile, bland, joyless, synthetic "cheese" thrives.

Modern mass-produced cheese throughout the region is now primarily made with cow's milk and synthetic rennet. Even small-scale cheesemakers in Lighvān now use imported Japanese rennet. This quickly made white block—"cheese by numbers" or "formula feta," as I like to call it—may resemble traditional cheese but lacks depth of flavor.

It is becoming fashionable to shop at supermarkets or large convenience stores and choose familiar brands over supporting small-scale producers. "Lighvān-style" cheese is increasingly produced by large companies. Consumers are wary of artisanal dairy, preferring factory-made, sterile, and "safe" alternatives.

However, I am encouraged by a few small companies and initiatives selling traditionally made dairy products, still made by villagers or nomads, which are promoted on social media platforms. Hopefully, small dairy shops like Mr. Abdi's and other independent producers will continue making and selling traditional cheese. One needs to be well connected or very determined to taste traditional *panir*.

Globalization has brought European-style cheese shops to large cities across Central Asia. These shops stock local cheese as well as imported cheese from around the world, including mold-ripened varieties. This shift shows how tastes are evolving, influenced by travel, social media, and cookery shows. I even found a company in Tatarstan making Camembert-style cheese.

Cheese is safe when consumed in moderation, though those on low-salt or low-fat diets need to be careful. Fresh cheese can be made and eaten without salt, but, as shown, salt is essential for both preservation and flavor.

To make "cheese," other coagulants, such as fig sap, nettles, and cardoons, are also used, though some can make the curds bitter. As we've seen in previous chapters, you can make "cheese" with plant-based milk and coagulants. These can also be inoculated with the same enzymes used in cheese making, allowing for varieties such as plant-based blue "cheese." Naturally, they have a different taste and texture. Some are quite delicious, but as I have said before, I wish these vegan pastes and blocks, usually nut based, were not called cheese.

I do not have the rennet recipe, so I can't give you a proper recipe, but I can suggest that you make a simple *panir* with shop-bought rennet or starter culture. Follow the instructions on the pack, which will tell you the quantity of milk and what temperature to heat the milk to.

Strain the curds and whey through a cloth-lined sieve and collect the whey.

Press the collected curds in the cloth thoroughly. This can be done by tightly packing the curds in a sieve lined with a few layers of cloth to extract the whey. Place a chopping board on top and weigh it down with heavy cans, a mortar, or something similar. Leave it for a couple of hours, changing the wet cloth as needed to remove as much liquid as possible from the curds.

In the meantime, if the rennet recipe didn't have any salt added to it, weigh the whey and make a four-to-one solution with salt. That is, if you have four cups of whey, add a cup of salt and mix to make sure it is fully dissolved.

Then unwrap the curds from the cloth and cut them to fit into a non-corrosive container. Also, you could cut them to the size you will eat over two to three days.

Put the curds into the container and top it up with the brining solution. Gently jiggle the container to free any air bubbles around the curds. Leave it in a cool, dark place for three months to "arrive," as we say. When it is ready to eat, continue to store it in brine. If ever your *panir* is too salty, you can take a piece out and put it in water for a few hours, changing the water a couple of times to reduce the salinity.

Among my family and friends in Iran, a favorite souvenir from the UK is my local artisanal cheddar. Over the years, I have taken hundreds of kilos of it. In recent years, I have also started taking Stichelton. You could call me a cheese mule.

10

Whey Ahead of You

Throughout the book I have promised to tell you more about the leftover liquid from fermenting dairy, variously called whey, buttermilk, *su, ayrān* or *doogh*. In this chapter, we look at some of the most common liquid dairy ferments of Central Asia. I will start with whey, *kumis*, and *arkhi*; then the region's most popular drink, *ayrān*; and last but by no means least, kefir. Generally, these products are consumed as beverages, and some, like *ayrān*, are also used as ingredients. Occasionally, these drinks were served as substitutes for water in harsh terrain or when water was not available nearby. The fact that they are full of nutrition and gut-friendly microorganisms was not understood at the time, but it may have been felt intuitively. People drank them and felt vigorous; today, science allows us to decipher and substantiate some of the health claims. They are embedded in the identity, culture, and hospitality of the peoples of Central Asia.

When I mention fermented drinks, some people in the West, depending on where they are from, might think of beer, wine, coffee, tea, kombucha, and increasingly, kvass and kefir. In Central Asia, we have grown up with white opaque fermented drinks. Those who are new to fermented dairy drinks might expect them to be mild and creamy—especially with names like buttermilk—but they are often surprised when they taste salty, fizzy, funky, or even slightly alcoholic.

These drinks are typically consumed cold, as thirst quenchers, paired with particular meals, and used as toasting libations in celebrations and rituals. Like all dairy products we have seen in the book, they go by various names across the regions. Generally, the more significant a dairy product is in the culture, the more varieties and subcategories it tends to have, reflecting its importance in daily life. For example, there are around thirty varieties of *kumis*. All of the drinks in this chapter are deeply rooted in the culture, signifying their importance in both diet and custom. In some cases, these customs were observed and adopted by those who came to the region, such as the British adding milk to tea.

Only a few years ago, these drinks were known only in Eurasia, and they were made at home for personal consumption—for example, using kefir grains to make kefir. Now they are widely available and can be purchased in high-street shops or online. Kefir is a great example; it is available in most supermarkets in the UK and has become a "superfood."

Let us dive straight into whey. If you have ever made butter, cheese, or strained yogurt, you will have noticed that a liquid comes out of it; this is whey. The color varies; yogurt whey is a light jade green, and in Āzari it is known as *joy su*, or green water.

In the previous chapters, I explained how dairy products such as *gooroot* or *gara gooroot* are made with whey, but they can also be consumed as a drink, added as an ingredient, used as a starter for other ferments, or fed to animals. Moreover, we have also seen that there are two types of whey: sweet whey, a by-product of churning fresh cream to make butter, and acid whey, a by-product of straining yogurt, making cultured butter, or curdling milk with an acidic element such as lemon juice.

Large amounts of whey are produced when dairy products are being made. Imagine the volume of whey generated by the dairy industry when making cheese, for example. In the past, industrial whey was often dumped into rivers, causing significant ecological harm. In the UK, starting in the mid-twentieth century, dairy producers were banned from this practice. This led them to repurpose the whey into a variety of innovative products. This highly nutritious liquid can be processed into preservable products that take up less room and are easier to transport. *Gara gooroot* is a great example, as it reduces huge quantities of whey into small amounts of preserves.

A modern method of dealing with large amounts of whey is to use "waste valorization" technology. It involves reusing, recycling, or composting waste materials to transform them into more useful products, such as new materials, chemicals, or energy sources. Such whey products are utilized in industrial applications such as biobased chemicals, biofuel, pharmaceuticals, animal feed, and plant fertilizer (Whey2Grow). In the food industry, whey is used to make protein powders, chocolate bars, fortified dairy products, liquors such as Irish cream, and as we saw in the previous chapter, whey butter. These are just a few of the applications.

At home, if you have whey, you can store it in the fridge for up to six months. I often keep a jar in my fridge, topping it up with various types of whey to use as needed. I have not frozen it myself, but it can also be frozen for later use. It can be enjoyed chilled, mixed into drinks, or further fermented into a Scandinavian whey wine known as *blaand*. It also works well in smoothies and cocktails for the adventurous mixologist among you.

I sometimes use whey as a starter for lacto-fermented vegetables, to soak beans or rice, or to sprout seeds. I often add a little whey to sauces, soups, or stews to thin them; in marinades; and at the end of cooking if a dish needs some moisture, seasoning, or acidity. I also add it to the cooking water for starches like pasta and potatoes; it adds salt and acid to the dishes.

It also works well as a liquid in baking, especially for quick soda bread, as the acid in whey interacts with the baking soda, raising the bread. Historically, before commercial leaveners like baking powder, whey was commonly used in baking. I usually make soda bread after making butter and use the whey as part of the liquid in the recipe. I know that heat kills some of the live bacteria, but other nutrients remain, and it is much better than just pouring it down the sink.

Outside the kitchen, I dilute whey with water and use it as a conditioner for acid-loving indoor and garden plants, such as blueberries and hydrangeas, or simply pour it onto the compost heap. Speaking of feeding, you can give whey to animals too. Apparently, it can also be used as a hair rinse, and in Tibet, a similar product is used for tanning leather.

Though high in macro- and micronutrients, whey is acidic and salty and should be used in moderation, especially when it is in concentrated form. Whey proteins have been particularly popular among weight lifters and strength trainers and are now increasingly used as a supplement for menopausal women. With growing interest in the role of protein and calcium in the aging process, whey is increasingly used in health foods and supplements.

Of all the drinks, *kumis*, as it is known in Turkic languages, and *airag* in Mongolian, is the one most revered by the people who make and drink it. There is much debate over its origins, who first made it, and who makes the best *kumis*. There is a story that suggests the city of Bishkek was named after the wooden device used to make *kumis*. The Kyrgyz people used a paddle called a *bishkek* to beat the milk.[1] I find this an interesting theory, and it could be true, but it is likely just a tale to assert a claim to *kumis*. Each region has its special *kumis*-making paraphernalia and considers its own techniques superior.

We are not sure where the name for *kumis*, variously written and pronounced, originates, but from the thirteenth century onward, *kumis* with different pronunciations became synonymous with fermented mare's milk in Turkic, Iranian, and Arabic languages. Helga Anetshofer, a Turkic-language specialist, suggests that *kumis* (which she Romanizes as *qimiz*) is a *wanderwort*, meaning "migrant word."

Frothy *kumis* is made from mare's milk, which some say is obtained in an extraordinary way. The natural sweetness of mare's milk lends itself well to this slightly alcoholic drink, which is white like milk. A similar version, made with camel or cow's milk, is called *shubat* or *chal*.

Kumis is primarily consumed in the central and northwestern regions of Eurasia, where wild horses are said to have first been domesticated and are still a significant part of life there. In this area, mares are a measure of wealth not only in financial terms but also in their provision of food, milk, and *kumis*. Some boast that while others have grape wine or grain beer, their mares eat grains or fruit to produce milk, making *kumis* a superior drink. The Mongol court collected four signature beverages from its territories: mead from Eastern Europe, wine from the Caucasus, rice beer from China, and the finest *kumis* from Mongolia.

As explained before, horses are highly adaptable, capable of sensing their environment and even human emotions. Archaeological finds such as saddles and bridles and their depictions in art, whether in wall paintings, carpets, or gold artifacts, illustrate their historical importance and close bond with humans.

Horses are strong and agile even when carrying a rider. They are social animals that like to roam and are well suited to the various terrains of Central Asian plains. Antelopes and horses have thrived in these environments, and research suggests that during the Mongol conquests, a favorable climate sustained a vast grazing area, allowing each soldier to have at least four horses. This abundance of horses enabled long-range travel and contributed to the development of exceptional horsewomanship and horsemanship, key factors in the success of the Scythian and, later, the Mongol Empires.

In my region, we do not have *kumis*, so I did not encounter it until later in life. It is mostly associated with Scythians, Kazakhs, Kyrgyz, Turkmen, and Mongols. *Kumis* and, similarly, *shubat* or *chal* are not just drinks; they have cultural significance, playing a vital role in daily life, celebrations, hospitality, and traditions of cultures across this vast region. It is nourishment, medicine, and identity. *Kumis* was so important in the cultures of these regions that a wife's skill in dairy processing, particularly *kumis* making, could determine her worth and, in some cases, even lead to divorce if she failed at it.

In the fifth century BC, the Greek historian Herodotus, in his *Histories*, vividly describes the Scythians' method of milking mares, which is both astonishing and peculiar. Based on his accounts, they used blind slaves who blew into the mares' vulvas to release milk, which was then frothed. In the thirteenth century, a Flemish Franciscan missionary and explorer, William of Rubruck, wrote that the Mongols valued *gara kumis*, which was made from black mare's milk, above all, possibly due to the rarity of black horses. He also remarked that "kumis makes the inner man most joyful." Other accounts, however, were not so favorable about the taste of *kumis*!

Mare's milk is not usually drunk raw. Besides being very high in lactose, it has a laxative effect, which is taken advantage of medicinally. Instead, it almost always gets fermented into *kumis*, which makes it more digestible. *Kumis* is made by lactic acid and alcohol fermentation. Traditionally prepared in springtime using special skin bags known as *saba* or in oak or maple barrels, a starter is added to the mare's milk. The process is labor intensive, requiring the ferment to be vigorously mixed initially and then stirred for shorter periods daily over two to three days.[2] Sometimes, the *saba* were tied to the yurt, and anybody going in or out shook them. Perhaps this is where the notion of all fermented dairy coming about by milk being shaken in a bag originates? It is then left to ferment for a few days, with daily gas release and stirring. Fizzy and volatile, *kumis* must be handled carefully. Experienced *kumis* makers can judge its readiness by aroma alone. It is typically consumed shortly after preparation, as its quality is said to deteriorate over time. If *kumis* is too strong or sour, it can be diluted with water or milk, though this is considered inferior.

Based on preparation methods, quality, and ripening time, *kumis* is divided into several types. There are many varieties of *kumis* that are categorized in different ways, but usually its age, the season in which it is made, and the alcohol level are considered. Each variety has a distinct flavor and texture. Some are valued more than others. In Kazakh, *kysyraktyn kumis* is made from the milk of a mare's first pregnancy, and *korabaly kumis* is stored for a few days and is then diluted with fresh mare's milk. There are also types distinguished by the winter diet of the mare, which was mainly grains, and this is known as winter *kumis*.

Kumis has its own serving paraphernalia besides the bags and barrels described above. There are special *kumis* bowls and ladles, some of which have one handle and two bowls. The first *kumis* of the season is given to guests, who then bless the owner's home.[3] The last *kumis* of the season has its own similar customs.

The best *kumis* is reserved for guests or used in rituals and celebrations. Foreign visitors have variously described its flavor as funky, sour, and even smoky. *Kumis* is traditionally served cold in a *piyala*, meaning a "small bowl," as a welcome drink. While it would be impolite to refuse it, it is acceptable to take a sip and pass the *piyala* back. Today, *kumis* is sold primarily in traditional cafés and is less popular with younger generations of the region.

Kumis is said to have many health benefits and is used in the region as part of traditional medicine to treat tuberculosis, gastritis, anemia, typhoid, cardiovascular diseases, and pancreatic problems, though these claims need to be studied further before being substantiated. Different types of *kumis* are given to the elderly, children, or new mothers to give them strength. Being slightly alcoholic, it is best consumed in moderation and avoided by those who are pregnant, driving, or riding a horse!

Raw mare's milk is hard to find. Industrial *kumis* is made by sweetening and fermenting cow's milk and can sometimes be bought online. Sweetened condensed milk is high in lactose and could be a good option for experimenting at home.

Chal and *shubat* are most commonly made in Kazakhstan, Turkmenistan, and Uzbekistan. A starter is added to the camel cow's milk, and it is left to sour; after a few days, it is stirred gently to take on a uniform, thick consistency and is quite rich. A two-day-old *shubat* is considered the best quality, and it is said, "the thicker the better." As with all the other drinks I discuss here, *shubat* and *chal* are assigned their own local beliefs and are drunk in the belief that they bestow health and vigor. They are said to be very good topically as well and are reputed to make maidens fair.

I'm not sure a plant-based, *kumis*-like drink is possible. The starter is animal-derived; however, if you don't mind that, you could try rice milk, which is high in sugars (though not lactose), and add the specific sugars or enzymes needed to mimic lactose.

Arkhi is a clear alcoholic spirit made by distilling the steam produced during the making of fermented milk products, or by extracting alcohol from whey. When *arkhi* is distilled several times, it becomes extremely concentrated and strong, so much so that it can burst into flames. In the West, it is sometimes referred to as "milk vodka." *Arkhi* is consumed by only some peoples of Central Asia, and its use has declined in predominantly Islamic regions. It is also known as *arkha*, *araka*, or *araga*, terms that sound similar to *arak*, *raki*, or *rakia*, which are names given to alcoholic drinks in European and Arabic languages.

Although *arkhi* is not a fermented dairy product per se but rather a by-product of dairy processing, I was so impressed by the ability to derive yet another product from milk that I chose to include it here. Turning steam from boiling whey into an alcoholic drink is another example of human resourcefulness and our ability to squeeze out every consumable drop from seemingly exhausted sources.

To make *arkhi*, whey is boiled in a large pot with a tight-fitting lid to first make *gooroot*, or *aarul* as Mongolians call it (it is mostly made and drunk by Mongolian people). A smaller pan is suspended inside the pot over the boiling liquid, and a concave lid is placed over the whole thing, with the curved surface toward the suspended pan. As the steam rises and condenses on the lid, it drips into the suspended pan. This clear liquid is *arkhi*. One type of *arkhi*, called Molochnaya Araga, is listed in the Ark of Taste and is made by the Tuva people in the Republic of Tyva.

Not quite the same thing, but while looking into this, I discovered that in the UK, there are several whey-based vodka- and gin-style drinks, such as Black Cow Vodka, Isle of Mull Spirit, and the famous Irish cream drinks. These are all made with leftover whey, mainly from cheese making, and the addition of other ingredients; for example, Isle of Mull uses homemade yeasts in their gin.

Western travelers in the thirteenth century reported that the Mongols drank alcohol but were punished if they became drunk. *Arkhi* was only consumed during special events and ceremonies, often used in toasts at weddings and religious observances but also to honor the blue sky, the earth, and their ancestors. They had learned distillation techniques on their "expeditions" to Iran, an example of knowledge exchange between the conqueror and the conquered.

In Turkic languages, there is a sour, salty drink known as *ayrān*; in Iranian languages, it is called *doogh*. It is also known as *tan, chalap,* and *chal,* and in parts of India, it is known as *lassi*. It is the most commonly consumed of the drinks in this chapter and my favorite. *Ayrān* is generally a salty drink, whereas *lassi* can be either sweet or salty and comes in a variety of flavors including a cannabis-flavored version. This beverage is typically served cold. Each type is made in a slightly different way, using different types of milk or yogurt, and each has its own unique flavor and place in national affections. In Āzari, *ayrān* and *doogh* are also synonyms for "whey" and are sometimes referred to as buttermilk in English.

When I am outside Turkic-language countries and I ask for *ayrān* (the Āzari name of the drink), they understand what I mean and bring what they know locally as *doogh*, or *chal*. They will not correct me. This is not due to laziness; rather, regional customs emphasize politeness and hospitality toward guests. As such, finding common ground is considered courteous. You see this elsewhere too: When nonnatives ask if *gooroot* is cheese, even though it is not, locals often just say yes out of courtesy. This explains the many references to cheese when discussing *gooroot* among other things in English-language sources.

Ayrān can be made with any type of milk, each contributing its own microorganisms, bringing a different flavor profile and mouthfeel to the final drink. Nowadays, it is a drink in its own right and is usually thick, but when it originated, it was likely quite diluted and may have emerged from the practice of rinsing out dairy-making containers with water.

Ayrān is an ancient savory drink that may have been safer and easier to find than clean water for pastoralists in the region. Iranian cookbook author Margaret Shaida writes that originally, milk was called *doogh*, and by the third and fourth centuries AD, the term came to describe a diluted yogurt drink. She goes on to explain that the verb "to milk an animal" is *dooshidan*. The past tense of this verb is the same as the Old Iranian word for "daughter," *dooghtarin*, meaning "those who milk animals," which became *dokhtar*, which is still used in Fārsi to mean "daughter." Through linguistic evolution, this influenced the word *daughter* in other languages. Interestingly, *pesar* (boy) in Iranian is from the old term *poost dar*, meaning "those who skin animals," possibly reflecting traditional gender roles in these societies, which I've touched upon in previous chapters.

Plutarch, a first-century AD Greek philosopher and historian, mentions an acidulated milk-like beverage consumed during the consecration of Persian kings 2,500 years ago. *Ayrān*-like drinks are also mentioned in the tenth-century Iranian epic *Shāhnāmeh* (*Shāhnāmeh: The Persian Book of Kings*) by Abolghāsem Ferdowsi, where shepherds are said to quench Ardeshir's (the founder of the Sasanian Empire) thirst with *doogh*,[4] and attributed to the fourteenth-century Shirazi satirist Boshaq al-Atameh, who said that "nobody admits their doogh is sour," meaning people never speak ill of themselves. As seen in the etymology of "daughter" and in various cultural artifacts and stories, dairy products have left lasting traces in our language and culture.

Ayrān shares many of the same health benefits and disadvantages as other fermented dairy products mentioned in this book. It is said that if you drink too much of the fizzy version, it could make you drowsy. This could be because it is heavy or perhaps because, in Āzarbāijān particularly, it is often consumed alongside rich foods such as *chillo chabāb* (skewered meat and buttery rice), or are they referring to mildly alcoholic *kumis* and kefir? I am a huge fan of the fizzy version of *doogh*, known as *Ob Ali*, which I bring back from Iran in cans. Predictably, it is also known as "yogurt soda" in America.

As to the varieties you can make at home, this depends on the milk or plain yogurt available to you. The simplest way to make *ayrān* is by diluting yogurt with water. It can be prepared with a single type of yogurt or a blend, especially if supplies are running low or you want to swill the yogurt pot out; there are no strict rules. You can use still or sparkling water and prepare it plain or flavored. *Ayrān* is typically served chilled. It is a nice nonalcoholic option for those who prefer unsweetened beverages.

Homemade *ayrān* flavored with rose and thyme

Ayrān

I like my *ayrān* sour, so I sour the yogurt before I make it, but you can make it by mixing still or sparkling water into plain yogurt, then adjusting the level of thickness and saltiness to your taste. Similarly, you can add flavorings such as rose petals, mint, or thyme.

Ingredients

2 tbsp. of plain soured yogurt
250 mL (9 fl. oz.) of either still or sparkling water

Mix the yogurt with 50 mL (1.7 fl. oz.) of the total water in the recipe until it is smooth. This prevents the need for straining later.

Pour the mixture into a tall glass and then top it up with the remaining water.

Stir well, taste, and adjust the salt as needed.

Optional extras: Enhance the flavor with herbs like mint or thyme, ideally in a liquid distilled form, as it blends in better. You can also add spices like chili or even curry leaves. If you use dried herbs, mix these into the initial base so they start to hydrate and blend rather than ending up floating on top.

Dairy-Free "Ayrān"

Use the recipe above, but with plant-based yogurt, then tweak it to your preferred flavor and dilution.

Mum's Fermented Ayrān

My mother regularly prepared fermented, flavorsome, fizzy *ayrān* at home, especially during the summer months. Her method is simple and flexible.

Ingredients

2 tbsp. of plain yogurt
120 mL (4 fl. oz.) of plain milk
1 tsp. salt

Add 2 tbsp. of plain yogurt and milk to a 2 L (4 pt.) container with a tight lid.

Add the salt and top up with water, leaving around 5 cm (2 inches) of space at the top of the bottle.

Leave this in a warm place around 25°C (77°F) for three days or until it becomes fizzy. It will take longer in colder climates.

Refrigerate the *ayrān* and serve it cold or with ice. Adjust the seasoning and flavorings when it is served; for example, use thyme or peppermint (dried or distilled).

A word of warning: Open the bottle over the sink, as sometimes it shoots out of the bottle!

Use the recipe above, substituting plant-based yogurt and milk, then tweak it to your preferred flavor and dilution.

To make sweet *lassi*, use fresh yogurt that has not been left to sour so that it is creamy and sweet. Typically, it is less diluted and mixed with milk instead of water; think of it as a fermented milkshake.

Lassi can be flavored with syrups, fruit purees, or sweet spices such as vanilla, cloves, or cardamom. As in the *ayrān* recipe, if using powdered dry spices, make a paste with a teaspoon of milk before adding them to the *lassi* to prevent them from floating on the surface.

Anba *Lassi*

Ingredients

2 tbsp. of either fresh or preserved mango puree, passed through a sieve to make sure it is smooth
2 tbsp. of full-fat fresh yogurt
100 mL (3.5 fl. oz.) of cold whole milk
⅛ tsp. of freshly ground cardamom seed

Mix the mango, yogurt, cardamom, and 2 tbsp. of milk until smooth. Whisk in the remaining cold milk. Serve over ice.

If you'd like it sweeter, add a teaspoon or more of grated jaggery (cane sugar).

Dairy-Free "*Lassi*"

This *lassi* can be made with plant-based milk and yogurt. I make it with mango puree, coconut milk, and yogurt and flavor it with kefir lime leaves and jaggery.

Another Central Asian method of creating fermented dairy is to use kefir grains, a mix of bacteria and yeast known as a SCOBY (symbiotic culture of bacteria and yeast). Unlike the other microorganisms discussed in this book, kefir grains are tangible. In my view, they are the empress of fermented products. The term *kefir* refers both to the fermented drink, similar to *ayrān*, and to the kefir grains themselves. Kefir grains need lactose, the sugar in milk, to survive. When kefir grains are added to milk in an environment between 5°C and 37°C (41°F and 99°F), the yeasts convert sugars into ethanol and carbon dioxide, making the resulting kefir fizzy, while the bacteria turn lactose into lactic acid, so it is slightly sour as well. Kefir contains a small amount of alcohol, but other bacteria present help break it down, further reducing its content.

I am in awe of kefir for many reasons, but primarily because it reproduces itself: The grains multiply, a magical and benevolent gift. They are referred to as grains, though they are not technically a grain; it is just a description. They are irregularly shaped, cream-colored small blobs that look like slimy cauliflower bits. I am really selling it, aren't I? Kefir has been treasured for millennia. Historically, kefir grains would have been another precious possession, a family heirloom to be looked after and shared.

Kefir is said to have originated in the Caucasus and was traditionally consumed in Central Asia. In Āzari, we pronounce it *chafir*, but in this book, I have used kefir, as it is more commonly known. Today, kefir is widely available as a ready-to-drink product, or you can buy the grains to make it yourself. With the increasing interest in the importance of our gut microbiota to our longevity and general well-being, kefir has become the supreme functional dairy ferment in some circles.

Kefir has the most diverse range of microbes (about three hundred different species that we know of so far). The resulting kefir product (I say "product," as it can be a drink or a thicker yogurt-like food eaten with a spoon) contains less lactose than yogurt and boasts a range of vitamins (B, C, A, and K) and minerals (calcium, magnesium, and potassium) and trillions of probiotics. Kefir is an extraordinary probiotic, perhaps one of the best sources of beneficial microorganisms to support gut health. When you add kefir grains to pasteurized, sterilized, or powdered milk, you reintroduce microorganisms that were removed during heat treatment, making the milk more nutritious and easier to digest.

The etymology of *kefir* traces back to Turkic origins. Some suggest it comes from *kef* in Turkish, pronounced *chef* in Āzari, meaning "pleasure," "enjoyment," or "drunkenness." Others trace it to *chöpür*, meaning "fizz" or "froth." I am more convinced by the latter definition and would add "to bloat" to these definitions as well, as all these terms describe what happens to the milk and the resulting kefir. With it being fizzy and acidic (dry), it is sometimes called "dairy champagne."

There are many stories about kefir's journey out of the Caucasus. My favorite is one in which the charms of a beautiful woman were used to deliver kefir grains to outsiders—in this case, slavs—and from there to the rest of the world. Today, kefir is consumed globally and is part of the "functional foods" category, a rapidly growing product range on supermarket shelves.

Today, genetic analysis allows us to understand its magic on a deeper level. The late evolutionary biologist Lynn Margulis beautifully described kefir:

> Although the kefir curd is a complex individual, a product of interacting aggregates of bacteria and fungi, it does not reproduce by sex. Rather, the kefir curd, which has no sex life, enlarges by direct growth, division, and death of its components. When mistreated by adverse environments, it disintegrates and dies. And, like any live individual, it never returns to life as that same individual.

She goes on to say, "Kefir can no more be made by the 'right mix' of chemicals or microbes than can oak trees or elephants."

We cannot make kefir grains, and they do not dissolve in liquid; if kept clean and handled properly, they multiply and pass on some of their properties to the substrate, growing about 5–7 percent per day When properly cared for, they remain stable and can last for generations. Some say that the grains we have today are descendants of the very first grains, a profound notion that makes me feel kindred to kefir.

The flavor and texture vary depending on the type of milk, environmental conditions (e.g., fridge or countertop), and how long you let it ferment. If left in a warm place or for a longer time, kefir becomes thicker, more sour, and fizzier. If refrigerated, it becomes milder, more fluid, and less fizzy.

Milk kefir can be fermented in an open or sealed container (aerobically or anaerobically). Traditionally, it was fermented aerobically under a cloth. I make mine in a jam jar covered with paper towels so that it isn't too fizzy. If contamination is a concern or you want it fizzier, you can seal the jar, but open it carefully.

Making kefir is an excellent way to begin your dairy ferment journey. You do not need specialized equipment or a specific milk type, though high-quality, full-fat milk is the best, but you do need to look after it. All you do is put the grains in milk, wait twelve to twenty-four hours, and then strain the ferment. You can use this straight away or store it in the fridge for a few days. Then you need to top the grains up with milk to restart the process. Keep going on like this, as a consistent routine seems to be key with kefir. You can slow fermentation by lowering the temperature or using fewer grains in more milk or speed it up by reversing these conditions.

To store or rest the grains, keep them in a large jar of milk in the fridge. They can also be frozen, though I haven't tried this myself. Kefir grains thrive with consistent care—that is, feed them milk, strain regularly, and maintain a stable environment. Like plants, they may "sulk" if conditions change, but they usually recover. I have personally experienced this attitude from my kefir grains. After giving them their first "bath," they seemed to sulk for a week, taking a fortnight to return to their usual vigor. In my attempts to coax them back, I began talking to them as I strained the kefir and added fresh milk, using a cheerful and encouraging tone, and I'd like to think this helped. I have friends who name their kombucha SCOBY and talk about it as if it were a pet, so I don't think this is too odd.

The fresher the kefir is, the better it is for you. This is one product I prefer to make rather than buy. After the grains consume the lactose, the mixture begins to separate into curds and whey. The ideal time to strain and enjoy the kefir is just before this occurs.

I have used a variety of milks in my kefir, such as nanny's, ewe's, and cow's milk. Use either semi-skimmed or full-fat, never skimmed, milk. I have not tried to make kefir with either raw milk or ultra-heat-treated (UHT) milk, which I avoid, but it can be done. Basically, you can use anything with lactose. When switching milk types, it may take a little while for the grains to adjust and settle into the new substrate. Raw milk may take a bit longer for the kefir to settle into, as it has many of its own bacteria, which will compete with kefir bacteria. However, eventually, it will start to ferment, and this in itself is enough to make the resulting kefir safe to consume.

Kefir can also be used as a starter to make *ayrān*, something similar to yogurt, or to clabber milk to make *lor*. By adding kefir to cream, you can create a *crème fraîche* or sour cream of sorts. If you then beat this cultured cream, you will make cultured butter and buttermilk. One can also use kefir in quick breads and other baked goods instead of buttermilk, but we must remember that heat destroys some of the benefits. I have also used it as a dressing and in cold soup.

To acquire your own kefir grains, consider reaching out through local social media networks or purchasing them, either hydrated or dehydrated. To rehydrate dried grains, place them in milk for one to three days, discarding the milk each day and starting with fresh milk until they are fully rehydrated and plumped up. Then start the fermentation process described above. Some people struggle to revive dehydrated grains, so it is better to get already hydrated ones.

Kefir is a nutritious food and one of the best sources of probiotics. Its diverse mix of bacteria and yeast can survive the harsh conditions of the gut, with significant benefits. Some of these include improved metabolic health, suppression of tumors, better wound healing, and the alleviation of allergies and asthma, just to name a few. However, most research on the effects of kefir, such as its impact on inflammation, thus far comes from laboratory and animal studies. More extensive human research is needed to substantiate these claims.

Kefir grains

Bowl of kefir

Kefir making

Making kefir

My *Chafir*

In my home, I make kefir at room temperature from November to March, when the average temperature in the kitchen is around 17°C (63°F). The rest of the year, I keep it in the fridge.

The following are proportions rather than ingredients.

   1 tbsp. of kefir grains
   1 L (2 pt.) of whole milk

Place kefir grains in a glass jar and pour milk over them.
   Cover with paper towels or cheesecloth and leave in a dark, draft-free spot at room temperature for 12-24 hours.
   When the milk curdles, strain to separate the grains from the kefir. The liquid is the kefir drink, and the grains remain in the sieve. Drink or refrigerate the kefir, and add fresh milk to the grains to start the next batch.

From now on, you can decide to continue making kefir at room temperature or in the fridge. Put the grains back in the same container (rinsed or not, depending on your preference).

If you prefer it thicker, more sour, or fizzier, pour the liquid back into the jar and leave it for a little longer.

Once you know how you like your kefir, you can establish a routine. Feed and strain it daily or every two to three days. The kefir will adapt to your schedule gradually.

When you make kefir yourself, you can control the level of sourness and thickness.

Sometimes, the kefir will separate into a thicker, opaque layer on top, with cloudy water below. This is an indication that your kefir is starting to overferment; it is still safe to consume, but it is past its peak. You can either mix it or skim off the opaque part (use it like yogurt) and drink the watery bit or use it as a starter for other ferments.

For a thicker kefir yogurt, strain the kefir product once it is sieved, or use a larger proportion of grains to milk.

If you neglect your kefir or change its routine, it might develop a yeasty taste. Simply wash the grains, restart the process, and be patient, as they will recover. I've made kefir with whole cow's, nanny, and ewe's milk, as well as with semi-skimmed milk, but I think the grains prefer whole milk. You can also make kefir with raw milk, as the bacteria and enzymes in the grains help eliminate pathogens naturally.

Kefir smells sour and yeasty, which is normal; however, if there is a rotten or unpleasant smell (you'll know) or if mold appears or the kefir or grains change in color, throw them away and start again with a new batch.

If you have more grains than you need, share them with friends and neighbors (this also acts like an insurance policy for you), eat them by blending them into the kefir or yogurt, or give them to your pets. Do not throw them away.

> ### Dairy-Free "Kefir"
>
> You can also make kefir with plant-based milk. For example, I once made almond milk kefir using 1 L (2 pt.) of the purest almond milk available and added 1 tbsp. of rinsed dairy kefir grains (rinsed in non-chlorinated water). I left it on the counter for two days as the grains became accustomed to the milk, then I started giving it about 50 mL (1.7 fl. oz.) of milk daily as it started to ferment. I was not that keen on the taste, but my vegan friend enjoyed it.
>
> I like coconut milk kefir. You can use fresh or desiccated coconut (blended into milk) or cartons or cans of coconut milk to do this.
>
> There is also water kefir, which I touched upon in chapter 3.

This is how I met my kefir grains. While researching for this book, I read about all the benefits of kefir and how easy it was to make and keep. I thought it would be useful to have firsthand experience and, in the process, treat my body. I asked around, and no one I knew had any. I posted on my local social media page asking for some and soon received a DM telling me to collect them whenever I was passing. It did not even occur to me that I was collecting something I was going to eat from a random stranger and that they could be harmful. I only realized this after a friend commented that it was trustworthy of me to do so. People who share food, especially ferment starters, are usually reliable. But if you are worried, buy kefir grains from a reputable source.

When I went to fetch the kefir, the door opened, and two people with rosy cheeks and glowing skin greeted me. I was invited into their kitchen, and—I am sure they won't mind my saying so—it was full of cats and cat paraphernalia. It turned out Helen was a naturopath, and she and her husband had been caring for this kefir for years. Moreover, they had both seen improvements in their health since taking it daily. She also kindly gave me a mini-lesson on how to keep the kefir going. She strained the cloudy liquid into a cup using a little sieve, then rinsed the grains and the jar they were in with tap water. After drying the jar, she put the grains back in, topped them with milk, and put the jar in the fridge. She said that when the grains grew, she threw them in the trash. She gave me about a tablespoon, or roughly ten grams (0.4 oz).

I came home that late May evening, placed the grains in a glass jar, topped them with pasteurized organic semi-skimmed milk, covered the jar with paper towels, and left it on the counter. After three days, I strained a bit and drank it cautiously. I had had kefir before and had not liked it, but this one was delicious. Since then, I have topped it up with the same small amount of milk daily, and I've started straining it every four days when it becomes sour and thicker.

From late June, I transferred the jar into the fridge, by which time the grains had grown, and I shared them with my sister and a school friend.

The yogurt we get in Tabriz is more sour than the natural yogurt we get here in Bath, and I wanted to substitute kefir for yogurt to eat with food rather than have it as a drink. In order to do this, I make mine with a small amount of milk and a lot of kefir grains, which I feed daily and strain every three days.

A few months after I got the kefir, I decided to give it a bath and wash the jar. By this time, the grains had grown to sixty grams (2 oz.) and were still in the same unwashed jar with its community of microorganisms. It felt a bit like turning over soil in the garden: I do it, but I also worry about disturbing all the good stuff. Anyway, I had to wash the jar, as it had become unsightly. I chose to use dechlorinated water to wash the kefir. To do this, I boiled two liters (4 pt.) of water and left it in a jug overnight to remove some of the chlorine. The next day, I bathed the grains in this water. I also washed the jar and restarted the process of topping it up and leaving it on the counter, as the weather was cooler by then. It took about a week for the grains to recover from this bath. I now bathe them twice a year and wash the jar every two months.

When I went to Iran in early 2025, I took half a teaspoon of grains for my mother. When I transferred them to a dish in her home and was pouring her home pasteurized milk, I was overcome with emotion. I welcomed the grains back to their homeland and hoped they'd like it there. I have never felt like this about anything else I have grown, cooked, or eaten. They quadrupled during my five-week visit, and I would like to think they were happy to be home.

A few days after I got back to the UK, my mother rang to say the kefir had changed flavor and was not as nice. I suggested she bathe it and then restart the process. She did this, and after a few days, it was OK. In a subsequent call, my mother teased me and asked if the kefir knew I had left and was missing me!

I enjoy caring for, preparing, eating, drinking, and sharing my kefir. If this inspires you, I encourage you to get some grains and give it a try. I feel a deep kinship with my kefir, which may descend from the same grains my ancestors once used. It is thrilling to realize that both the kefir grains and I come from the same part of the world, a place that has made such significant contributions to world history, culture, and food.

Much of the region's oral and embodied knowledge is fragile yet vital, and it must be preserved. Exploring and writing about foodways—particularly the stories of these fermented dairy traditions—is about more than the food itself; it is about honoring and keeping alive knowledge, traditions, and practices. It has brought me immense joy to share these discoveries with you.

Acknowledgments

I would like to begin by thanking Janet Joyce and the team at Equinox Publishing Ltd. for suggesting I write this book. I am grateful to Brianna Blackburn and her team for their editorial guidance. I also thank Cateryn Kilgarriff, who originally commissioned me to write a book about *gooroot*, a project that ultimately led to this work.

I appreciate the encouragement and camaraderie of my fellow Oxford Food Symposium symposiasts, particularly my dear friends Gamze Ineceli, Aglaia Kremezi, Elisabeth Luard, Claudia Roden, and Anna Del Conte. A special thank you to Mary Margaret Chappell for her friendship and counsel when I needed it most.

I felt fortunate to be able to consult Charles Perry and Harold McGee; thank you both for your advice.

Thanks also to Nilufar Kuchkarova for photos and stories from Uzbekistan; Dr. Āblimit Baki Elterish for sharing Uyghur traditions; Ulan Iskakov for his knowledge of Kazakh food and hospitality; Dr. Jessica Hendy for reviewing archaeological data; and Dr. Gökçen Durukoğlu for translating her article into English.

I am indebted to the many women who, though they preferred not to be photographed or named, shared their time, patience, and knowledge. Special thanks to Raviyeh Khānum; to Minā and Sorayā Jalilzadeh for the *gooroot* disks and my first taste of real milk; to Messrs. Sadeghiān and Abdi for their knowledge and delicious dairy; to Helen for sharing her kefir grains, some of which have now returned to Āzarbāijān; and especially to my patient, wise, and constant companion, Sālehe Sālehpour-Oskoui, whom I call مامان.

Finally, thank you to Frank Davis for his support throughout this process.

Notes

Introduction

1 *Foodways* neatly captures the intersection of food with culture, tradition, and history—that is, the eating habits, culinary practices, attitudes, beliefs, and behaviors that shape how people engage with food in daily life.
2 Ingredient lists only include what the company is legally bound to declare.
3 The Ark of Taste is the world's largest catalog of cultural and traditional biodiversity related to food and agriculture. "Ark of Taste," Slow Food, https://www.slowfood.com/biodiversity-programs/ark-of-taste/.

1. Origins

1 Bronowski, *Ascent of Man*.
2 "Maasai Aborigines Steal Food."
3 A term used by James Henry Breasted in 1914. Breasted, *Ancient Times*, 101.

2. Udder Beginnings

1 Swigart, "Rarest Tuscan Cheese."
2 Liles, "Does Photo Show Baby?"
3 To help you remember the number of humps on the Bactrian and Dromedary camels, rotate the first letter of each word ninety degrees counterclockwise.
4 These animals are also referred to as the Przewalski's or Dzungarian horse.

5　A mule is the offspring of a jackass and a mare. A hinny is the offspring of a male horse and a female donkey.
6　Tengrism is a pagan tradition practiced in Central Asia.
7　Nowruz is the first day of the year for cultures that consider the exact moment of the vernal equinox in the Northern Hemisphere as the beginning of the year.
8　Cows are now given vaccines to help reduce their flatulence in an attempt to reduce global warming.
9　This method was developed by Professor Leighton of Sloan Water Technology Ltd. (https://www.sloanwatertechnology.co.uk).
10　In fresh whole milk, the fat globules naturally rise to form the cream layer. Heating the milk makes the fat rise more quickly. The proteins in milk are normally distributed throughout, but exposure to air at the surface causes them to become distorted, clump together, and eventually form a continuous skin. Heating the milk makes this skin formation faster. In an email to me, Harold McGee provides the following explanation: Because both processes happen simultaneously, the thickened surface layer is a combination of cream and protein. Their relative proportions will depend on the milk composition and the processing temperatures and times.

　　If the thickened layer is removed and the process is repeated, the relative proportions of protein and fat will change, since some of each has been removed in earlier steps.
11　Wilson, "Woe of Wasted Milk."

3. Fermentation, a Microbial Marvel

1　Bacteria and yeasts, known for their adaptability, thrived in diverse conditions, and as Richard Dawkins says in *The Selfish Gene*, their genes are selfish and want to survive.
2　No reference to the term *yak* could be identified in the available sources on Turkic languages; instead, terms such as *yak* or *Tibetan ox* appear to be used. In Fārsi, the word *ghazh* is defined in various ways but is generally understood to mean "to crawl."
3　Quoted in F. Guarner et al., "Should Yoghurt Cultures Be Considered Probiotic?"
4　The Ordinary Cook, "Herman—the Friendship Cake."

5. Which Came First, the Yogurt or the Pot?

1 *Laban* is the Arabic word for yogurt, and *labna* is strained yogurt.

6. *Gooroot*: Dried Dairy Balls

1 The "Delia Effect" was coined in 1998. It can be defined as the high-profile recommendation (usually by a celebrity) of a product that in turn results in overnight success.
2 The Afshari tribes are among the oldest Turkic tribes. The Afshari sheep, characterized by their golden black fleece, small heads in proportion to their large, woolly bodies, and large droopy ears, are one of the many fat-tailed sheep breeds found in western Iran. Their wool, meat, and milk are of high quality.
3 Perry, "Cheese of the Pharaohs."
4 Caproic acid is an oily liquid with a smell that is similar to that of goats.
5 Foda et al., "Antimicrobial Activity of Dried Spearmint."
6 "WHO Studies Reveal."

7. A Sour Sidekick

1 Perry, "Horseback Kitchen."

8. Spinning, Laughing, Dancing to Her Favorite Song

1 This is part of British folklore relating to butter; however, the bright-yellow reflection is more about attracting pollinators.
2 Kearns, "Irish Bog Butter."
3 Bourdain, *Tony Bourdain Boxset*, 76.
4 It is called clotted cream because the cream rises and thickens, or "clots," on the surface when heated.

9. A Separation: *Su & Dan*

1. In private conversation with Ivan Patrick Day. In the UK in times past, "the general theory was that the stomach was like a cauldron and the astringent quality of cheese would help seal the cauldron and allow the digestion to carry out its work."

10. Whey Ahead of You

1. More plausibly, the city's name comes from a term meaning "five peaks" or "place beyond the mountains."
2. Despite my research, I could not find a reliable, accurate, and well-documented starter culture recipe for *kumis*. It seems you may have to create your own using some store-bought or gifted *kumis* or using backslopping in either mare's or camel's milk. It will require a lot of work, patience, and trial and error.
3. This tradition, a spring festive rite of Kazakh horse breeders, has been nominated as an Intangible Cultural Heritage to UNESCO. "Traditional Spring Festive Rites of the Kazakh Horse Breeders—UNESCO Intangible Cultural Heritage," n.d., https://ich.unesco.org/en/RL/traditional-spring-festive-rites-of-the-kazakh-horse-breeders-01402.
4. Here, *doogh* could have been whey, milk, or *doogh*; we do not know. Batmanglij, "Milk and Its By-Products."

Bibliography

Albala, Ken. "Milk: Nutritious and Dangerous." In *Oxford Symposium Proceedings: Milk: Beyond the Dairy*, edited by Harlan Walker, 19–30. Prospect Books, 2000.

Alberta Milk. "Why Is Milk White?" 2025. https://albertamilk.com/ask-dairy-farmer/why-is-milk-white/#:~:text=While%20milk%20is%20primarily%20composed,makes%20the%20liquid%20appear%20white.

Algar, Ayla. "Bushaq of Shiraz: Poet, Parasite and Gastronome." *Petits Propos Culinaires* 31 (1989): 9.

Allsen, Thomas T. *Culture and Conquest in Mongol Eurasia*. Cambridge University Press, 2008.

American Institute of Chemical Engineers. "Waste Valorization." https://www.aiche.org/topics/energy/waste-valorization#:~:text=Waste%20valorization%20is%20the%20process,or%20other%20sources%20of%20energy.

Anetshofer, Helga. "The Turkic Word *Qımız* 'Fermented Mare's Milk': Early Historical Textual Evidence and Origin." *Zemin* 6 (2023): 34–81. https://doi.org/10.5281/zenodo.10435940.

Armitage, Susie. "Make the Ancient Road Snack of Central Asian Nomads: Qurt Is Salty, Long-Lasting, and Packed with Protein." *Atlas Obscura*, March 8, 2021. https://www.atlasobscura.com/articles/what-is-qurt.

Asher, David. *The Art of Natural Cheesemaking: Using Traditional, Non-Industrial Methods and Raw Ingredients to Make the World's Best Cheeses*. Chelsea Green, 2015.

Aubaile-Sallenave, Françoise. "'Al-Kishk': The Past and Present of a Complex Culinary Practice." In *A Taste of Thyme: Culinary Cultures of the Middle East*, edited by Sami Zubaida and Richard Tapper, 105–41. Tauris Park, 2000.

Avery, Martha. *Women of Mongolia*. Avery Press, 1996.
Ball, Warwick. *The Eurasian Steppe: People, Movement, Ideas*. Edinburgh University Press, 2021.
Batmanglij, Najmieh. "Milk and Its By-Products in Ancient Persia and Modern Iran." In *Oxford Symposium Proceedings: Milk: Beyond the Dairy*, edited by Harlan Walker, 64–74. Prospect Books, 2000.
Baumgarthuber, Christine. *Fermented Foods: The History and Science of a Microbiological Wonder*. Reaktion Books, 2021.
Bell, L., et al. "The Impact of a Walnut-Rich Breakfast on Cognitive Performance and Brain Activity Throughout the Day in Healthy Young Adults: A Crossover Intervention Trial." *Food & Function* 16, no. 5 (3 Mar. 2025): 1696–1707. doi:10.1039/d4fo04832f.
Bellwood, Peter. "The Beginnings of Agriculture in Southwest Asia." In *First Farmers: The Origins of Agricultural Societies*. Blackwell, 2005.
Berdugo, Sophie. "Before the Stone Age: Were the First Tools Made from Plants Not Rocks?" *New Scientist*, 5 November 2024.
Blackmon, Susan. *The Meme Machine*. Oxford University Press, 2000.
Bonner, Michael J. "Nomad Riders of the Steppe: Ancient Rome May Have Flourished Only Because the Huns and Other Tribes Did Not Turn West Sooner." *Dorchester Review*, Spring/Summer 2019, 92–96.
Bostock, John, and Henry T. Riley. *The Natural History. Pliny the Elder*. Taylor and Francis, 1855.
Bourdain, Anthony. *Tony Bourdain Boxset: Kitchen Confidential & Medium Raw*. A & C Black, 2011.
Breasted, James Henry. *Ancient Times: A History of the Early World*. Boston: Ginn and Company, 1916.
Brigand, Robin, and Olivier Weller. *Archaeology of Salt: Approaching an Invisible Past*. Sidestone Press, 2015.
Bronowski, Jacob. *The Ascent of Man*. BBC, 1973.
Bryant, K. L., C. Hansen, and E. E. Hecht. "Fermentation Technology as a Driver of Human Brain Expansion." *Communications Biology* 6 (2023): article 1190. https://doi.org/10.1038/s42003-023-05517-3.
Buchanan, Dominic, Wayne Martindale, Ehab Romeih, and Essam Hebishy. "Recent Advances in Whey Processing and Valorisation: Technological and Environmental Perspectives." *Dairy Technology* 76, no. 2 (2023): 291–312. https://doi.org/10.1111/1471-0307.12935.

Çağatay, Nilufer. *Social and Economic History of the Turkic Peoples*. Türk Kültürünü Araştırma Enstitüsü, 1978.

Calahorrano-Moreno, Micaela Belan, et al. "Contaminants in the Cow's Milk We Consume? Pasteurization and Other Technologies in the Elimination of Contaminants." *F1000 Research* 11 (2022): 91. doi:10.12688/f1000research.108779.1.

Campbell-Thomson, Olga. "Kurt (Kurut)." Cabinet. Accessed May 20, 2023. https://www.cabinet.ox.ac.uk/kurt-kurut-0.

Cappers, R. T. J. *Digital Atlas of Traditional Food Made from Cereals and Milk*. University of Groningen Library, 2018.

Chong, Ann Qi, Siew Wen Lau, Nyuk Ling Chin, Rosnita A. Talib, and Roseliza Kadir Basha. "Fermented Beverage Benefits: A Comprehensive Review and Comparison of Kombucha and Kefir Microbiome." *Microorganisms* 11, no. 5 (2023): 1344. https://doi.org/10.3390/microorganisms11051344.

Clauson, Gerard. *An Etymological Dictionary of Pre-Thirteenth-Century Turkish*. Oxford University Press, 1972. https://doi.org/10.1002/ajhb.10156.

Coleman, Leo, ed. "Introduction." In *Food: Ethnographic Encounters*, 1–16. Oxford: Berg, 2011.

Collins, Andrew. *Karahan Tepe: Civilization of the Anunnaki and the Cosmic Origins of the Serpent of Eden*. Rochester, VT: Bear & Company, 2024.

Costa Louzada, M. L., et al. "Ultra-Processed Foods and the Nutritional Dietary Profile in Brazil." *Revista de Saúde Pública* 49 (2015): 38. https://doi.org/10.1590/S0034-8910.2015049006132.

Craze, Paul. "Early Human Evolution and the Skulls of Dmanisi." *Significance* 10, no. 6 (2013): 6–11. https://doi.org/10.1111/j.1740-9713.2013.00703.x.

Dailymotion. "Maasai Aborigines Steal Food from Cheetahs and Lions, Can They Do It!" April 4, 2022. Accessed November 5, 2024. https://www.dailymotion.com/video/x89yckw.

Dalrymple, William. "The Silk Road Still Casts a Spell, but Was the Ancient Trading Route Just a Western Invention?" *The Guardian*, October 6, 2024.

da Silva Ghizi, Angela Camila, et al. "Kefir Improves Blood Parameters and Reduces Cardiovascular Risks in Patients with Metabolic Syndrome."

Pharma Nutrition 16 (2021): 2213–4344. https://doi.org/10.1016/j.phanu.2021.100266.

Davis, Clara M. "Results of the Self-Selection of Diets by Young Children." *Canadian Medical Association Journal* 41, no. 3 (1939).

Dawkins, Richard. *The Selfish Gene*. Oxford University Press, 2006.

Debusmann, Bernd, Jr. "Texas Dairy Farm Explosion Kills 18,000 Cows." BBC News, April 13, 2023. https://www.bbc.co.uk/news/world-us-canada-65258108.

Değhirmencioğhlu, Nurcan, et al. "Influence of Tarhana Herb (*Echinophora sibthorpiana*) on Fermentation of Tarhana, Turkish Traditional Fermented Food." *Food Technology and Biotechnology* 43, no. 2 (April 2005): 175–79. https://www.researchgate.net/publication/230669249_Influence_of_Tarhana_Herb_Echinophora_sibthorpiana_on_Fermentation_of_Tarhana_Turkish_Traditional_Fermented_Food.

Der Haroutunian, A. *The Yogurt Cookbook*. Grub Street, 2010.

Domínguez-Rodrigo, Manuel, and Travis Rayne Pickering. "Early Hominid Hunting and Scavenging: A Zooarcheological Review." *Evolutionary Anthropology: Issues* 12, no. 6 (2003): 275–82.

Dudd, S. N., and R. P. Evershed. "Direct Demonstration of Milk as an Element of Archaeological Economies." *Science* 282 (1998): 1478–81.

Elliott, Stuart, dir. *Irresistible: Why We Can Not Stop Eating*. BBC TV, 2024.

Erickson, Britt E. "Acid Whey: Is the Waste Product an Untapped Goldmine? Companies, Food Scientists Develop Innovative Solutions to Handle Tons of Greek Yogurt Byproduct." *Chemical & Engineering News*, February 6, 2017. https://cen.acs.org/articles/95/i6/Acid-whey-waste-product-untapped.html.

Evans, Miranda, et al. "Detection of Dairy Products from Multiple Taxa in Late Neolithic Pottery from Poland: An Integrated Biomolecular Approach." *Royal Society Open Science* 10 (2023). https://doi.org/10.1098/rsos.230124.

Evershed, R., et al. "Earliest Date for Milk Use in the Near East and Southeastern Europe Linked to Cattle Herding." *Nature* 455 (2008): 528–31. https://doi.org/10.1038/nature07180.

Farhadi, Asghar. *Jodāi-e Nāder az Simin*. Culver City, CA: Sony Pictures Classics, 2011.

Ferdowsi, A. *Shahnameh: The Persian Book of Kings*. Trans. Dick Davis. Penguin Classics, 2006.

Ferring, Reid, et al. "Earliest Human Occupations at Dmanisi (Georgian Caucasus) Dated to 1.85–1.78 Ma." *Proceedings of the National Academy of Sciences of the United States of America* 108, no. 26 (2011): 10432–36. https://doi.org/10.1073/pnas.1106638108.

Fisberg, Mauro, and Rachel Machado. "History of Yogurt and Current Patterns of Consumption." *Nutrition Reviews* 73, no. 1 (2015): 4–7. https://doi.org/10.1093/nutrit/nuv020.

Foda, Mervat I., M. A. El-Sayed, N. A. El-Sayed, and M. A. Amer. "Antimicrobial Activity of Dried Spearmint and Its Extracts for Use as White Cheese Preservatives." *Alexandria Journal of Food Science and Technology* 6, no. 1 (2009): 55–65.

Food and Agriculture Organization of the United Nations. *Codex Alimentarius: Milk and Milk Products*. 2nd ed. FAO, 2011. http://www.fao.org/docrep/015/i2085e/i2085e00.pdf.

Fox, Patrick F., Timothy P. Guinee, Timothy M. Cogan, and Paul L. H. McSweeney. *Fundamentals of Cheese Science*. Springer, 2000.

Glendinning, John I. "Is the Bitter Rejection Response Always Adaptive?" *Physiology & Behavior* 56 (1994): 1217–27.

Golubeva, Raisa Kurt. "A Precious Stone." Written in the twentieth century, exact date unknown.

Gouin, Paul. "Rapes, jarres et faisselles: La production et l'exportation des produits laitiers dans l'Indus du 3ᵉ millénaire." *Paléorient* 16, no. 2 (1990): 37–54. https://www.persee.fr/doc/paleo_0153-9345_1990_num_16_2_4531.

Green, Nile, ed. *The Persianate World: The Frontiers of a Eurasian Lingua Franca*. 1st ed. University of California Press, 2019.

Guarner, F., G. Perdigon, G. Corthier, S. Salminen, B. Koletzko, and L. Morelli. "Should Yoghurt Cultures Be Considered Probiotic?" *British Journal of Nutrition* 93, no. 6 (2005): 783–86.

Güler, Zehra. "Changes in Salted Yoghurt During Storage." *International Journal of Food Science and Technology* 42, no. 2 (2007): 235–45. https://doi.org/10.1111/j.1365-2621.2006.01505.x.

Hansen, Valerie. *The Silk Road: A New History*. Oxford University Press, 2012.

Hawkes, K., James, F. O'Connel, Nicholas G. Blurton Jones, Helen Alvarez, and Eric L. Charnov. "The Grandmother Hypothesis and Human Evolution." In *Adaptation and Human Behaviour: An Anthropological*

Perspective, edited by L. Cronk, N. Chagnon, and W. Irons, 237–58. Aldine de Gruyter, 2003.

Hendy, Jessica, et al. "Ancient Proteins from Ceramic Vessels at Çatalhöyük West Reveal the Hidden Cuisine of Early Farmers." *Nature Communications* 9 (2018): article 4064. https://doi.org/10.1038/s41467-018-06335-6.

Herodotus. *Herodotus: The Histories*. Penguin, 1996.

Homer. *The Odyssey*. Translated by E. V. Rieu and D. C. H. Rieu. Penguin Classics, 2009.

Ibrahim, S. A., et al. "Fermented Foods and Probiotics: An Approach to Lactose Intolerance." *Journal of Dairy Research* 88, no. 3 (2021): 357–65. https://doi.org/10.1017/S0022029921000625.

Illikoud, Nassima, Marine Mantel, Malvyne Rolli-Derkinderen, Valérie Gagnaire, and Gwénaël Jan. "Dairy Starters and Fermented Dairy Products Modulate Gut Mucosal Immunity." *Immunology Letters* 251–52 (2022): 91–102. https://doi.org/10.1016/j.imlet.2022.11.002.

Issayeva, K. S., and G. T. Kazhibayeva. *Technology of National Food Production in Kazakhstan and Central Asia*. Academic Council of S. Toraighyrov Pavlodar State University, 2018.

Itan, Y., Adam Powell, Mark A. Beaumont, Joachim Burger, and Mark G. Thomas. "The Origins of Lactase Persistence in Europe." *Nature* 436, no. 7054 (2010): 1348–51. doi:10.1371/journal.pcbi.1000491.

Ji, Jing, Weilin Jin, Shuang-Jiang Liu, Zuoyi Jiao, and Xiangkai Li. "Probiotics, Prebiotics, and Postbiotics in Health and Disease." *MedComm* 4, no. 6 (2020). https://doi.org/10.1002/mco2.420.

Jin, Liya, Fahu Chen, Carrie Morrill, Bette L. Otto-Bliesner, and Nan Rosenbloom. "Causes of Early Holocene Desertification in Arid Central Asia." *Climate Dynamics* 38 (2012): 1577–91. https://doi.org/10.1007/s00382-011-1086-1.

Jones, Nora. "Seven Years." Track 2 on *Come Away with Me*. 20th Anniversary Super Deluxe Edition. Capitol Records, 2002.

Kabukcu, Ceren, et al. "Cooking in Caves: Palaeolithic Carbonised Plant Food Remains from Franchthi and Shanidar." *Antiquity* 97, no 391 (2023): 12–28. https://doi.org/10.15184/aqy.2022.143.

Katz, Sandor E. *The Art of Fermentation*. Chelsea Green, 2012.

Katz, Sandor E. *Sandor Katz's Fermentation Journeys: Recipes, Techniques, and Traditions from Around the World*. Chelsea Green, 2021.

Katz, Sandor E. *Wild Fermentation: The Flavor, Nutrition, and Craft of Live-Culture Foods*. Chicago Review Press, 2016.

Kearns, D. "Irish Bog Butter Proven to Be '3500 Years' Past Its Best Before Date." News and Opinion, University College Dublin, March 14, 2019, Accessed November 5, 2024. https://www.ucd.ie/newsandopinion/news/2019/march/14/irishbogbutterproventobe3500yearspastitsbestbeforedate/.

Khosrova, Elaine. *Butter: A Rich History*. Algonquin Books, 2016.

Knight, Chris, James R. Hurford, and Michael Studdert-Kennedy, eds. *The Evolutionary Emergence of Language: Social Function and Origins of Linguistic Form*. Cambridge University Press, 2000. https://catdir.loc.gov/catdir/samples/cam031/00020471.pdf.

Kretchmer, Norman. "Expression of Lactase During Development." *American Journal of Human Genetics* 45, no. 4 (1989): 487–88.

Kurlansky, Mark. *Milk! A 10,000-Year Food Fracas*. Bloomsbury, 2018.

Langley, Michelle C., and Thomas Suddendorf. "Mobile Containers in Human Cognitive Evolution Studies: Understudied and Underrepresented." *Evolutionary Anthropology* 29, no. 6 (2020): 299–309. https://doi.org/10.1002/evan.21857.

Lauson, G. *An Etymological Dictionary of Pre-Thirteenth-Century Turkish*. Oxford University Press, 1972. https://doi.org/10.1002/ajhb.10156.

Levinson, Stephen C., and Judith Holler. "The Origin of Human Multi-Modal Communication." *Philosophical Transactions of the Royal Society of London Series B, Biological Sciences* 369, no. 1651 (2014). https://doi.org/10.1098/rstb.2013.0302.

Levi-Strauss, Claude. "The Culinary Triangle." 1966. In *Food and Culture: A Reader*, edited by Carole Counihan and Penny van Esterik, 36–43. Routledge, 2008.

Liles, Jordan. "Does Photo Show Baby Feeding from Goat's Udders in 1927?" Snopes, February 16, 2022. Accessed October 9, 2023. https://www.snopes.com/fact-check/baby-feeding-goat-photo/.

Linford, Jenny. *The Missing Ingredient: The Curious Role of Time in Food and Flavour*. Particular Books, 2018.

Mandela, Nelson. *Long Walk to Freedom*. Flash Point / Roaring Brook Press, 2009.

Margulis, Lynn. "Did Sex Emerge from Cannibalism? Sex, Death and Kefir." *Scientific American*, November 23, 2011. Accessed November 5,

2024. https://www.scientificamerican.com/article/sex-death-kefir-lynn-margulis/.

McGee, Harold. *On Food and Cooking: The Science and Lore of the Kitchen*. Scribner, 2004.

McLendon, Russell. "If Your Fridge Dies, Should You Put Frogs in Your Milk?" Treehugger, February 10, 2021. Accessed January 2024. https://www.treehugger.com/if-your-fridge-dies-should-you-put-frogs-in-your-milk-4868045.

Melletti, Mario. "Cattle Domestication: From Aurochs to Cow." *Fifteen Eighty Four*, February 18, 2016. https://cambridgeblog.org/2016/02/cattle-domestication-from-aurochs-to-cow/.

Michaylova, Michaela, Svetlana Minkova, Katsunori Kimura, Takashi Sasaki, and Kakuhei Isawa. "Isolation and Characterization of *Lactobacillus delbrueckii* ssp. *bulgaricus* and *Streptococcus thermophilus* from Plants in Bulgaria." *FEMS Microbiology Letters* 269, no. 1 (2007): 160–69. https://doi.org/10.1111/j.1574-6968.2007.00631.x.

Minervino, Antonio Humberto Hamad, et al. "*Bubalus bubalis*: A Short Story." *Frontiers in Veterinary Science* 7 (2020). https://doi.org/10.3389/fvets.2020.570413.

Moʿīn, M. in ʿAlīzadeh, ʿAzīzo ʾllah (ed.). فرهنگ فارسی معین, (Farhang-e Moʿīn). *Encyclopedic Dictionary*. Vol. 2, p. 1231. 3rd ed. Ādnā, 2002.

Monty Python. *Monty Python's Life of Brian*. London: Eyre Methuen, 1979.

Neef, Reinder. "Overlooking the Steppe-Forest: A Preliminary Report on the Botanical Remains from Early Neolithic Göbekli Tepe (Southeastern Turkey)." *Neo-Lithics* 2 (2003): 13–19. https://www.exoriente.org/repository/NEO-LITHICS/NEO-LITHICS_2003_2.pdf.

Nicosia, Fabrizio Domenico, et al. "Plant Milk-Clotting Enzymes for Cheesemaking." *Foods* 11, no. 6 (2022): 871. https://doi.org/10.3390/foods11060871.

Niimura, Yoshihito, and Masatoshi Nei. "Evolution of Olfactory Receptor Genes in the Human Genome." *Proceedings of the National Academy of Sciences of the United States of America* 100, no. 21 (2003): 12235–40. https://doi.org/10.1073/pnas.1635157100.

Olatide, M., et al. "Pilot Study on Chilli Stalks as a Source of Non-Dairy Lactic Acid Bacteria in Yogurt Making." *Applied Food Science Journal* 3, no. 1 (2019): 5–8.

The Ordinary Cook. "Herman—the Friendship Cake." January 19, 2012. Accessed October 10, 2024. https://theordinarycook.co.uk/2012/01/19/herman-the-friendship-cake-2/.

Oskenbey, Moldir. "Fermented Dairy Products in Central Asia: Methods for Making Kazakh *Qurt* and Their Health Benefits." *Crossroads* 14 (2016). https://ostasien-verlag.de/zeitschriften/crossroads/cr/pdf/CR_14_2016_205-218_Oskenbay.pdf.

Ostadrahimi, Alireza, et al. "Effect of Probiotic Fermented Milk (Kefir) on Glycemic Control and Lipid Profile in Type 2 Diabetic Patients: A Randomized Double-Blind Placebo-Controlled Clinical Trial." *Iranian Journal of Public Health* 44, no. 2 (2015): 228–37.

Perry, Charles. "Cheese of the Pharaohs." *Los Angeles Times*, July 14, 1999. Accessed November 2022.

Perry, Charles. "Dried, Frozen and Rotted: Food Preservation in Central Asia and Siberia." In *Oxford Symposium Proceedings: Cured, Fermented and Smoked Foods*, edited by Helen Saberi, 240–48. Prospect Books, 2010.

Perry, Charles. "The Horseback Kitchen of Central Asia." In *Food on the Move: Proceedings of the Oxford Symposium on Food and Cookery 1996*, edited by Harlan Walker, 243–48. Prospect Books, 1997.

Perry, Charles. "Note from Lady Carolyn Conran." *Petits Propos Culinaires* 55 (1997): 34.

Perry, Charles. "Tracta/Trahanas/Kishk." *Petits Propos Culinaires* 14 (1983): 58.

Peterson, V. Spike. "Sex Matters: A Queer History of Hierarchies." *International Feminist Journal of Politics* 16, no. 3 (2014): 389–409. https://doi.org/10.1080/14616742.2014.913384.

Plutarch. *Plutarch Lives Aratus, Artaxerxes, Galba, Otho Volume XI*. Harvard University Press, 1994.

Pollan, Michael. *Cooked: A Natural History of Transformation*. Penguin, 2014.

Potts, Olivia. *Butter: A Celebration—a Joyous Immersion in All Things Butter*. Headline Home, 2022.

Pruetz, J. D., and N. M. Herzog. "Savanna Chimpanzees at Fongoli, Senegal, Navigate a Fire Landscape." Supplement, *Current Anthropology* 58, no. S16 (2017): S337–S350. https://doi.org/10.1086/692112.

Rappaport, Roy. *Ritual and Religion in the Making of Humanity*. Cambridge University Press, 1999.

Rezai-Ghassemi, Simi. "From the Steppe to Space: Portable Power of Qurut." In *Oxford Symposium Proceedings: Portable Food*, edited by Mark McWilliams, 306–11. Prospect Books, 2023.

Rosenstock, Eva, Julia Ebert, and Alisa Scheibner. "Cultured Milk: Fermented Dairy Foods Along the Southwest Asian–European Neolithic Trajectory." Supplement, *Current Anthropology* 62, no. S24 (2001). https://www.journals.uchicago.edu/doi/full/10.1086/714961.

Sagan, C. *Cosmos*. 1st ed. Ballantine Books, 1985.

Saladino, Dan. "Why Luxury Cheese Is Being Targeted by Black Market Criminals." BBC News, November 10. https://www.bbc.co.uk/news/articles/crmz42pjpnjo. Accessed November 10, 2024.

Salque, M., et al. "Earliest Evidence for Cheese Making in the Sixth Millennium BC in Northern Europe." *Nature* 493 (2013): 522–25. https://doi.org/10.1038/nature11698.

Sandoiu, Ana. "What Too Much Salt Can Do to Your Brain." Medical News Today, January 16, 2018. https://www.medicalnewstoday.com/articles/320612. Accessed September 2023.

Savaiano, Dennis A., and Robert W. Hutkins. "Yogurt, Cultured Fermented Milk, and Health: A Systematic Review." *Nutrition Reviews* 79, no. 5 (2021): 599–614.

Science Daily. "Ancient Humans Brought Bottle Gourds to the Americas from Asia." December 14, 2005. Accessed June 20, 2025. https://www.sciencedaily.com/releases/2005/12/051214081513.htm.

Scott, Ashley, et al. "Emergence and Intensification of Dairying in the Caucasus and Eurasian Steppes." *Nature Ecology & Evolution* 6, no. 6 (2022): 813–22. https://doi.org/10.1038/s41559-022-01701-6.

Settanni, L., and G. Moschetti. "Non-Starter Lactic Acid Bacteria Used to Improve Cheese Quality and Provide Health Benefits." *Food Microbiology* 27, no. 6 (2010): 691–97. https://doi.org/10.1016/j.fm.2010.05.023.

Shaida, Margaret. "Yoghurt in Iran." In *Oxford Symposium Proceedings: Milk: Beyond the Dairy*, edited by Harlan Walker, 309–15. Prospect Books, 2000.

Sherratt, Andrew. "The Secondary Exploitation of Animals in the Old World." *World Archaeology* 15, no. 1 (1983): 90–104.

Singh, Pranav K., and Nagendra P. Shah. "Other Fermented Dairy Products: Kefir and Koumiss." In *Yogurt in Health and Disease Prevention*, edited by Nagendra Shah, 87–106. Academic Press, 2017.

Smith, Natasha. "Bovaer Cow Feed Additive Explained." Food Standards Agency, December 5, 2024. Accessed December 10, 2024. https://food.blog.gov.uk/2024/12/05/bovaer-cow-feed-additive-explained/.

Soedamah-Muthu, S. S., and Janette de Goede. "Dairy Consumption and Cardiometabolic Diseases: Systematic Review and Updated Meta-Analyses of Prospective Cohort Studies." *Current Nutritional Reports* 7 (2018): 171–82. https://doi.org/10.1007/s13668-018-0253-y.

Spirits of Mongolia. "Mongolian Traditional Distilled Alcohol." February 16, 2022. https://spiritsofmongolia.com/mongolian-traditional-distilled-alcohol/.

Spuler, Bertold. *History of the Mongols*. Translated by Helga and Stuart Drummond. University of California Press, 1972.

Staubwasser, M., and H. Weiss. "Holocene Climate and Cultural Evolution in Late Prehistoric–Early Historic West Asia." *Quaternary Research* 66 (2006): 372–87. https://doi.org/10.1016/j.yqres.2006.09.001.

Suresha, Ron J. *Extraordinary Adventures of Mullah Nasruddin*. 2nd ed. CreateSpace, 2018.

Suryanarayan, Akshyeta, et al. "Lipid Residues in Pottery from the Indus Civilisation in Northwest India." *Journal of Archaeological Science* 125 (2021). https://doi.org/10.1016/j.jas.2020.105291.

Swallow, Dallas M. "Genetics of Lactase Persistence and Lactose Intolerance." *Annual Review of Genetics* 37, no. 1 (2003): 197–219. https://doi.org/10.1146/annurev.genet.37.110801.143820.

Swigart, Rob. "The Rarest Tuscan Cheese." Life in Italy, January 23, 2017. Accessed February 2024. https://lifeinitaly.com/rarest-tuscan-cheese/.

Syer Cuming, H. "On Ancient Sieves and Colanders." *Journal of the British Archaeological Association* 25, no. 3 (1869): 244–50. doi.org/10.1080/00681288.1869.11887546.

Tavakoli, H. R., M. Aghazadeh Meshgi, N. Jonaidi Jafari, M. Izadi, R. Ranjbar, and A. A. Imani Fooladi. "A Survey of Traditional Iranian Food Products for Contamination with Toxigenic *Clostridium botulinum*." *Journal of Infection and Public Health* 2, no. 2 (2009). https://doi.org/10.1016/j.jiph.2009.03.001.

Trask, Robert Lawrence. *The Dictionary of Historical and Comparative Linguistics*. Edinburgh University Press, 2022.

Trinkaus, Erik, and Fereidoun Biglari. "Middle Paleolithic Human Remains from Bisitun Cave, Iran." *Paléorient* 3, no. 2 (2006): 105–11. https://doi.org/10.1016/j.jhevol.2019.102643.

University College London. "Humans Struggle to Differentiate Imagination from Reality." April 21, 2023. Accessed March 3, 2024. https://www.ucl.ac.uk/news/2023/apr/humans-struggle-differentiate-imagination-reality.

Uteuova, Aliya. "Qurt: A Kazakh 'Cheese of Resilience.'" BBC, April 27, 2021. Accessed January 2, 2023. https://www.bbc.com/travel/article/20210426-qurt-a-kazakh-cheese-of-resilience.

Vanaeon, Elkin. "Cheese from Egypt." elkinvanaeon.net, 2013. Accessed April 14, 2013. http://elkinvanaeon.net/Alchemy/Cultured/Mish-Cheese.htm.

Van Tulleken, Chris. *Irresistible: Whey We Can Not Stop Eating*. BBC TV, Producer Naomi Pallas. Broadcast November 25, 2024. Accessed November 25, 2024.

Wikipedia. "Burning Man." Accessed August 27, 2022. https://en.wikipedia.org/wiki/Burning_Man.

Wikipedia. "William of Rubruck." July 29, 2019. Accessed October 10, 2025. https://en.wikipedia.org/wiki/William_of_Rubruck.

Wilkin, S., et al. "Dairy Pastoralism Sustained Eastern Eurasian Steppe Populations for 5,000 Years." *Nature Ecology & Evolution* 4 (2020): 346–55. https://doi.org/10.1038/s41559-020-1120-y.

Wilkin, S., et al. "Dairying Enabled Early Bronze Age Yamnaya Steppe Expansions." *Nature* 598 (2021): 629–33. https://doi.org/10.1038/s41586-021-03798-4.

Wilson, Bee. *The Secret of Cooking: Recipes for an Easier Life in the Kitchen*. 4th Estate, 2023.

Wilson, David. "The Woe of Wasted Milk." Sustainable Food Trust, March 22, 2023. Accessed November 5, 2024. https://sustainablefoodtrust.org/news-views/the-woe-of-wasted-milk/#:~:text=We%20waste%20an%20estimated%20490,the%20face%20for%20the%20farmer.

World Health Organization. "WHO Studies Reveal Kazakhstan Has Among the Highest Levels of Salt Intake Globally." March 7, 2019. https://www.who.int/europe/news/item/07-03-2019-who-studies-reveal-kazakhstan-has-among-the-highest-levels-of-salt-intake-globally.

Wrangham, Richard W., James Holland Jones, Greg Laden, David Pilbeam, and NancyLou Conklin-Brittain. "The Raw and the Stolen: Cooking and the Ecology of Human Origins." *Current Anthropology* 40 (1999): 567–94.

Wu, Xiaohonu, et al. "Early Pottery at 20,000 Years Ago in Xianrendong Cave, China." *Science* 366 (2012): 1696–1701. https://doi.org/10.1126/science.1218643.

Xygalatas, Dimitris. *Ritual: How Seemingly Senseless Acts Make Life Worth Living*. Profile Books, 2022.

Yang, Yimin, et al. "Proteomics Evidence for Kefir Dairy in Early Bronze Age China." *Journal of Archaeological Science* 45 (2014): 178–86. https://doi.org/10.1016/j.jas.2014.02.005.

Žajlybay, Galym. *The Black Shawl*. Poem-requiem, translation by N. Černova, 2014. In *Kurt*, Olga Campbell-Thomson. Accessed January 5, 2024. https://www.cabinet.ox.ac.uk/kurt-kurut-0.

Zhukovskaya, Natalia. "The Milk Food of the Mongolian-Speaking Nomads of Eurasia in a Historical and Cultural Perspective." *Acta Ethnographica Hungarica* 53, no. 2 (2009): 307–14. https://doi.org/10.1556/AEthn.53.2008.2.5.

Zilhao, Joao, et al. "Symbolic Use of Marine Shells and Mineral Pigments by Iberian Neandertals." *The Proceedings of the National Academy of Sciences* 107, no. 3 (2020): 1023–28.

Index

A

aarul 130
   See also *gooroot*
Abdi, Mr. (Tabriz dairyman) 7, 38, 119, 122, 147
abomasum 48, 117
āghārty
   as guest offering 33
   See also *bánbhia*; white food
Akmolinsk Camp 92
al-Jahiz 22
amasi 76
anaerobic fermentation 49
Anetshofer, Helga 127
   See also *wanderwort*
animally funk 8, 86
animals
   domestication 18, 20, 27
   milk-producing female animals 6, 26–29, 51–52
   roles in pastoralism 8, 114
A1/A2 milk genes 37, 51
arkhi 130–31
ayrān
   buttermilk 124–25, 131–36
   recipes 133–35
   See also *doogh*; *su*
Āzarbāijān 1–2, 25, 28
   See also Iran
Āzari language
   humor 29
   libricide and suppression 3
   oral tradition 3
Āzerbaijān 2–3

B

backslopping 68–69, 114
Bahman, Mr. 74–75
bāl gaymākh 110–11
bánbhia 7
Beauty of Xiaohe 21
Bishkek 127
Blackman, Susan 22
bokashi 59
Bonner, Michael 1
boolamā 25
borāni 68
Bourdain, Anthony 103
brain development 8, 11–14, 17, 113
brining 118, 121, 123
buffalo 21, 29–30
   milk 52, 110
butter 38, 63, 64, 79, 99–109, 112–14
   cultured 100, 130
   dish 107, 112
   recipe 104–7
   sweet 100
   whey butter 99, 104, 112, 118–19

C

camel cow 24
   milk 27, 52, 127, 129

Central Asia 1–2
   climatic zones 7–8
   mammals' dairy heritage 25, 32–33
chal 128, 130
   See also *shubat*
Changiz Khan 69, 94
cheddar 115–16, 118, 123
cheese
   brined cheese 119–20
   cultural significance 113
   curd separation 21, 48, 50, 53, 117–18
   plant-based alternatives 123
   See also *panir*; rennet
chortān 21, 62, 73, 77–81, 87–93
churn (*nehra*, *yāyik*, *mashk*) 100–102
clabber
   animal stomach 27, 117
   containers 15–17, 43, 78
   definition 46–47, 63–64, 78
   gourd 16, 44, 60, 76
   lor and *chortān* 79, 114, 118, 138
Coleman, Leo 5
colostrum 24–25
   See also *boolamā*
cow 24
   milk 52
   modern cows 34–37

D

dairying
   allergies 53, 57–59
   archaeological evidence of 19–20, 114
   dairy intolerance 57–59
   modern practices 33–40
   See also lactase persistence
dan 113
Darwin, Charles 22
deadstock 18
domestication
   of animals for dairy 18–20, 27
   horses 8, 27, 127
   See also horses; mares
donkeys 24, 28–29
   cultural humor 29
   See also jenny
doogh 131–32
   See also *ayrān*
dri (female yak) 24, 30
   milk 52
   See also yak

E

Eurasia 1–2, 26–28, 32–33
   migrations across 12–13, 27, 32–33
ewes 6, 24–26, 51

F

female animals
   recognition of 24–25, 52
fermentation
   biochemical process 45–46
   container link 16, 44
   controlled 46
   early fermented foods and evolution 13–15
   mixed 48–49
   natural occurrence and origins 13–14, 27
   spontaneous or wild 46
fire 17, 19, 42
frog peptides 38
functional foods 58, 69, 137

G

gala 65
Galen 66
gara gooroot 95–97, 125
gātikh 68
gaymākh 30, 99–111

Index

ghee 99, 108–9
   See also *sāri yāgh* or *sāri māi*
Göbekli Tepe 19
Golubeva, Raisa 92
gooroot
   definition and preparation 4–6, 79–87, 88–89, 91, 94–95
   etymology and regional spellings 4
   gooroot ali 84, 86
   historical uses 77–79
gut-brain axis 55–56

H

Harappan civilization 21
Herman friendship cake starter 60
Herodotus 128
horses
   domestication and spiritual role 7, 8, 27–28, 114, 127–29
   See also mares

I

industrial dairy
   mechanization and waste 34–37, 40–41
   milk components 50–51, 60
Iran
   Āzari population and language 2–3
   commerce 38, 82, 85, 122
   geography 2–3, 12

J

jenny 24, 28
   milk 52
   Pule 115
Jones, Nora 100

K

Karahan Tepe 19

kashk 1, 80, 82, 88
   kashke siāh 96
   kashkmāl 84
   kashksāb 85
kefir 5–6, 42–43
   etymology 136
   grains (SCOBY) 43, 48–49, 135–39
   health benefits 136, 139
   making kefir 137–38, 141–44
   plant-based kefir 143
Khanum, Raviyeh 95, 147
Kollath, Werner 55
kumis 28, 32, 127–29
   airag 9, 126
   cultural significance 127–28
   preparation 128–29
   qimiz 127
   varieties 129–30

L

lactase persistence 19–20, 57–58
lactic acid bacteria 47–48, 64, 67, 118
lactose 45, 47–48, 50
   lactose allergy 58
   lactose intolerance 8, 19, 57–59
Līghvān (village) 31, 51, 62
   panir 118, 122
Linford, Jenny 89
lor 79, 88, 114, 117–18

M

Mandela, Nelson 76
mares 24, 51
   milk 28, 32, 50–52, 127–29
Margulis, Lynn 137
McGee, Harold 46, 88
Mead, Margaret 22
mesophilic fermentation 49
microorganisms 42–49, 59, 120

milk
  cultural respect, ritual, and symbolic meaning 7, 32–33, 35
  industrial production 9, 34–37
  raw milk 36, 38, 62–63, 91, 111, 138
  use of milk by humans 6–9, 15, 18
  waste and sustainability 40–41
  See also *sut* and *sheer*
Mongols 6, 9, 126–31
  milk rituals and seasonal diets 22, 28, 32
  See also *aarul*; horses

N
nanny goat 6, 19–21, 24, 27
  milk 21, 27, 32, 37, 52
nomadism 6, 8, 20, 22–23, 32–33, 77, 122

P
panir 48, 87, 113–14, 118–23
  chat and *chup paniri* 122
Perry, Charles 82, 96
plant-based milk 39–40, 52–53, 72–73, 92, 134–35, 143
Platais, Gertruda 92
Pliny the Elder 99
Plutarch 132
postbiotics 55
pottery 20–21
prebiotics 55
probiotics 55–56, 69
proteins and lipid analysis 20–21

R
raw milk 36, 38
  home pasteurizing 36–37
red food 22
rennet 47–48, 115–18, 120–23

rituals
  dairy 32–33
  fire 18–19
Russel Wallace, Alfred 22

S
Saadi 68
Sadaghian, Mr. 80–81
Sahand, Mount 31, 118
salt 61, 68
sāri yāgh or *sāri māi* 108–9
SCOBY (symbiotic culture of bacteria and yeast) 48, 135
Shaida, Margaret 68, 131–32
sheep 24, 27, 51
  fat-tailed 108
shells as containers 15–16
Sherratt, Andrew 18
shubat 52, 127–30
Sir Mix-a-Lot 51
Slow Food Foundation, Ark of Taste 9, 116, 130
sow 24, 26
spontaneous fermentation 46, 64, 114, 116
  See also wild fermentation
starter cultures 46, 48–49, 72, 125–26, 138
su 32, 113, 119
sut and *sheer* 32
suzma 72, 79

T
Tabriz 29, 36, 38, 80, 82, 103
Texas dairy explosion 35
thermophilic fermentation 49
tool use 13–14
traditional medicine
  Unani 22, 88
  Unani, Ayurvedic, and Chinese 33
transhumance 23

U

UHT milk 35, 41
   making kefir 138
   making yogurt 70–71
UNESCO Intangible Cultural Heritage 9

W

wanderwort 127
water kefir (*tibicos*) 48–49, 143
whey 21, 48, 51, 79, 82, 117–19, 123–26
   butter 99
   gara gooroot 95
   gooroot 79
   sweet and acid whey 125
   whey products 125
   whey valorization 125
   See also *ayrān*; *doogh*; *su*
white food 7, 22
   See also *āghārty*; *bánbhia*
wild fermentation 46, 64
William of Rubruck 128
Wilson, Bee 89

wolves 18, 26
women 6–7, 79, 100
Wrangham, Richard 17
   See also fire

Y

yak 24, 26–27, 29
   See also dri
yeasts 8, 42–43, 46–50, 135–36
   in SCOBY 48, 135
   yeasty kefir 142
yogurt 7, 27, 45, 65–80, 89–92, 96–97, 104, 108
   dried yogurt 21, 44, 95, 97
   health 56
   industrial yogurt production 70
   See also *gātikh*; *suzma*
yurt 6, 102, 114, 128

Z

Žajlybay, Galym 94
zymology/zymurgy 44

www.ingramcontent.com/pod-product-compliance
Lightning Source LLC
Chambersburg PA
CBHW060819190426
43197CB00038B/2109